It is fascinating to review the diverse forecasts that experts have made in recent years about China's politics, economy, environment, and international relations. How is it possible that they disagree so much? Irvine wisely counsels caution, insisting that we must allow critical uncertainties to temper our ambition to know the future.

Andrew J. Nathan, *Class of 1919 Professor of Political Science, Columbia University*

In all the discussion over the implications of China's rise, one of the knottiest questions is just what sort of a China is it going to be. Roger Irvine offers a unique and path-breaking contribution to the task of forecasting China's role, with special attention paid to the relationship between China experts and the uncertainty that is intrinsic to this challenging but necessary enterprise. His balancing of theory with practical case studies of prediction across a variety of fields is exemplary, and his conclusions are judicious and well supported by the evidence he presents.

Richard Rigby, *Professorial Fellow, Crawford School of Public Policy, Australian National University*

Forecasting China's Future

China's future development is likely to have a huge impact on twenty-first century global outcomes. It is therefore surprising that, thus far, so little attention has been given to comparing and evaluating expert forecasts of China's future in the post-Mao era.

This book presents an illuminating and comprehensive summary record of contrasting and competing expert forecasts and judgements about the major issues confronting China within four principal domains – political, economic, environmental, and international. After considering the principal forecasting methods available to experts, the author comments critically on the degree of success achieved in using those methods and emphasises the confusion created by the polarisation of opinion and by the failure of many experts to accept the high degree of uncertainty that characterises most of the key issues. The book recommends a new approach based on the study of a hierarchy of critical uncertainties and on continuing analysis of opposing expert opinions about these uncertainties. It emphasises the potential for both positive and negative outcomes for these critical uncertainties, and the importance of maximising the potential for positive outcomes through improved analytical and policy frameworks.

Providing insights for specialists and non-specialists into the most critical issues that will determine China's future direction, this book will be of particular interest to students and scholars of political, economic, environmental, and international relations issues in China and Asia, as well as to readers in business and government.

Roger Irvine is a Visiting Research Fellow at the University of Adelaide, South Australia. He was a visiting scholar at Tsinghua University, China during 2010 and at National Chengchi University, Taiwan during 2014.

Routledge Contemporary China Series

1 **Nationalism, Democracy and National Integration in China**
Leong Liew and Wang Shaoguang

2 **Hong Kong's Tortuous Democratization**
A comparative analysis
Ming Sing

3 **China's Business Reforms**
Institutional challenges in a globalised economy
Edited by Russell Smyth, On Kit Tam, Malcolm Warner and Cherrie Zhu

4 **Challenges for China's Development**
An enterprise perspective
Edited by David H. Brown and Alasdair MacBean

5 **New Crime in China**
Public order and human rights
Ron Keith and Zhiqiu Lin

6 **Non-Governmental Organizations in Contemporary China**
Paving the way to civil society?
Qiusha Ma

7 **Globalization and the Chinese City**
Fulong Wu

8 **The Politics of China's Accession to the World Trade Organization**
The dragon goes global
Hui Feng

9 **Narrating China**
Jia Pingwa and his fictional world
Yiyan Wang

10 **Sex, Science and Morality in China**
Joanne McMillan

11 **Politics in China Since 1949**
Legitimizing authoritarian rule
Robert Weatherley

12 **International Human Resource Management in Chinese Multinationals**
Jie Shen and Vincent Edwards

13 **Unemployment in China**
Economy, human resources and labour markets
Edited by Grace Lee and Malcolm Warner

14 **China and Africa**
Engagement and compromise
Ian Taylor

15 **Gender and Education in China**
Gender discourses and women's schooling in the early twentieth century
Paul J. Bailey

16 **SARS**
Reception and interpretation in three Chinese cities
Edited by Deborah Davis and Helen Siu

17 **Human Security and the Chinese State**
Historical transformations and the modern quest for sovereignty
Robert E. Bedeski

18 **Gender and Work in Urban China**
Women workers of the unlucky generation
Liu Jieyu

19 **China's State Enterprise Reform**
From Marx to the market
John Hassard, Jackie Sheehan, Meixiang Zhou, Jane Terpstra-Tong and Jonathan Morris

20 **Cultural Heritage Management in China**
Preserving the cities of the Pearl River Delta
Edited by Hilary du Cros and Yok-shiu F. Lee

21 **Paying for Progress**
Public finance, human welfare and inequality in china
Edited by Vivienne Shue and Christine Wong

22 **China's Foreign Trade Policy**
The new constituencies
Edited by Ka Zeng

23 **Hong Kong, China**
Learning to belong to a nation
Gordon Mathews, Tai-lok Lui, and Eric Kit-wai Ma

24 **China Turns to Multilateralism**
Foreign policy and regional security
Edited by Guoguang Wu and Helen Lansdowne

25 **Tourism and Tibetan Culture in Transition**
A place called Shangrila
Åshild Kolås

26 **China's Emerging Cities**
The making of new urbanism
Edited by Fulong Wu

27 **China–US Relations Transformed**
Perceptions and strategic interactions
Edited by Suisheng Zhao

28 **The Chinese Party-State in the 21st Century**
Adaptation and the reinvention of legitimacy
Edited by André Laliberté and Marc Lanteigne

29 **Political Change in Macao**
Sonny Shiu-Hing Lo

30 **China's Energy Geopolitics**
The Shanghai Cooperation Organization and Central Asia
Thrassy N. Marketos

31 **Regime Legitimacy in Contemporary China**
Institutional change and stability
Edited by Thomas Heberer and Gunter Schubert

32 **US–China Relations**
China policy on Capitol Hill
Tao Xie

33 **Chinese Kinship**
Contemporary anthropological perspectives
Edited by Susanne Brandtstädter and Gonçalo D. Santos

34 **Politics and Government in Hong Kong**
Crisis under Chinese sovereignty
Edited by Ming Sing

35 **Rethinking Chinese Popular Culture**
Cannibalizations of the canon
Edited by Carlos Rojas and Eileen Cheng-yin Chow

36 **Institutional Balancing in the Asia Pacific**
Economic interdependence and China's rise
Kai He

37 **Rent Seeking in China**
Edited by Tak-Wing Ngo and Yongping Wu

38 **China, Xinjiang and Central Asia**
History, transition and crossborder interaction into the 21st century
Edited by Colin Mackerras and Michael Clarke

39 **Intellectual Property Rights in China**
Politics of piracy, trade and protection
Gordon Cheung

40 **Developing China**
Land, politics and social conditions
George C. S. Lin

41 **State and Society Responses to Social Welfare Needs in China**
Serving the people
Edited by Jonathan Schwartz and Shawn Shieh

42 **Gay and Lesbian Subculture in Urban China**
Loretta Wing Wah Ho

43 **The Politics of Heritage Tourism in China**
A view from Lijiang
Xiaobo Su and Peggy Teo

44 **Suicide and Justice**
A Chinese perspective
Wu Fei

45 **Management Training and Development in China**
Educating managers in a globalized economy
Edited by Malcolm Warner and Keith Goodall

46 **Patron-Client Politics and Elections in Hong Kong**
Bruce Kam-kwan Kwong

47 **Chinese Family Business and the Equal Inheritance System**
Unravelling the myth
Victor Zheng

48 **Reconciling State, Market and Civil Society in China**
The long march towards prosperity
Paolo Urio

49 **Innovation in China**
The Chinese software industry
Shang-Ling Jui

50 **Mobility, Migration and the Chinese Scientific Research System**
Koen Jonkers

51 **Chinese Film Stars**
Edited by Mary Farquhar and Yingjin Zhang

52 **Chinese Male Homosexualities**
Memba, *Tongzhi* and Golden Boy
Travis S. K. Kong

53 **Industrialisation and Rural Livelihoods in China**
Agricultural processing in Sichuan
Susanne Lingohr-Wolf

54 **Law, Policy and Practice on China's Periphery**
Selective adaptation and institutional capacity
Pitman B. Potter

55 **China–Africa Development Relations**
Edited by Christopher M. Dent

56 **Neoliberalism and Culture in China and Hong Kong**
The countdown of time
Hai Ren

57 **China's Higher Education Reform and Internationalisation**
Edited by Janette Ryan

58 **Law, Wealth and Power in China**
Commercial law reforms in context
Edited by John Garrick

59 **Religion in Contemporary China**
Revitalization and innovation
Edited by Adam Yuet Chau

60 **Consumer-Citizens of China**
The role of foreign brands in the imagined future china
Kelly Tian and Lily Dong

61 **The Chinese Communist Party and China's Capitalist Revolution**
The political impact of the market
Lance L. P. Gore

62 **China's Homeless Generation**
Voices from the veterans of the Chinese civil war, 1940s–1990s
Joshua Fan

63 **In Search of China's Development Model**
Beyond the Beijing consensus
Edited by S. Philip Hsu, Suisheng Zhao and Yu-Shan Wu

64 **Xinjiang and China's Rise in Central Asia, 1949–2009**
A history
Michael E. Clarke

65 **Trade Unions in China**
The challenge of labour unrest
Tim Pringle

66 **China's Changing Workplace**
Dynamism, diversity and disparity
Edited by Peter Sheldon, Sunghoon Kim, Yiqiong Li and Malcolm Warner

67 **Leisure and Power in Urban China**
Everyday life in a medium-sized Chinese city
Unn Målfrid H. Rolandsen

68 **China, Oil and Global Politics**
Philip Andrews-Speed and Roland Dannreuther

69 **Education Reform in China**
Edited by Janette Ryan

70 **Social Policy and Migration in China**
Lida Fan

71 **China's One Child Policy and Multiple Caregiving**
Raising little Suns in Xiamen
Esther C. L. Goh

72 **Politics and Markets in Rural China**
Edited by Björn Alpermann

73 **China's New Underclass**
Paid domestic labour
Xinying Hu

74 **Poverty and Development in China**
Alternative approaches to poverty assessment
Lu Caizhen

75 **International Governance and Regimes**
A Chinese perspective
Peter Kien-Hong Yu

76 **HIV/AIDS in China – The Economic and Social Determinants**
Dylan Sutherland and Jennifer Y. J. Hsu

77 **Looking for Work in Post-Socialist China**
Governance, active job seekers and the new Chinese labor market
Feng Xu

78 **Sino-Latin American Relations**
Edited by K.C. Fung and Alicia Garcia-Herrero

79 **Mao's China and the Sino-Soviet Split**
Ideological dilemma
Mingjiang Li

80 **Law and Policy for China's Market Socialism**
Edited by John Garrick

81 **China–Taiwan Relations in a Global Context**
Taiwan's foreign policy and relations
Edited by C. X. George Wei

82 **The Chinese Transformation of Corporate Culture**
Colin S. C. Hawes

83 **Mapping Media in China**
Region, province, locality
Edited by Wanning Sun and Jenny Chio

84 **China, the West and the Myth of New Public Management**
Neoliberalism and its discontents
Paolo Urio

102 Islam and China's Hong Kong
Ethnic identity, Muslim networks and the new Silk Road
Wai-Yip Ho

103 International Regimes in China
Domestic implementation of the international fisheries agreements
Gianluca Ferraro

104 Rural Migrants in Urban China
Enclaves and transient urbanism
Fulong Wu, Fangzhu Zhang and Chris Webster

105 State-Led Privatization in China
The politics of economic reform
Jin Zeng

106 China's Supreme Court
Ronald C. Keith, Zhiqiu Lin and Shumei Hou

107 Queer Sinophone Cultures
Howard Chiang and Ari Larissa Heinrich

108 New Confucianism in Twenty-First Century China
The construction of a discourse
Jesús Solé-Farràs

109 Christian Values in Communist China
Gerda Wielander

110 China and Global Trade Governance
China's first decade in the World Trade Organization
Edited by Ka Zeng and Wei Liang

111 The China Model and Global Political Economy
Comparison, impact, and interaction
Ming Wan

112 Chinese Middle Classes
China, Taiwan, Macao and Hong Kong
Edited by Hsin-Huang Michael Hsiao

113 Economy Hotels in China
A glocalized innovative hospitality sector
Songshan Sam Huang and Xuhua Michael Sun

114 The Uyghur Lobby
Global networks, coalitions and strategies of the World Uyghur Congress
Yu-Wen Chen

115 Housing Inequality in Chinese Cities
Edited by Youqin Huang and Si-ming Li

116 Transforming Chinese Cities
Edited by Mark Y. Wang, Pookong Kee and Jia Gao

117 Popular Media, Social Emotion and Public Discourse in Contemporary China
Shuyu Kong

118 Globalization and Public Sector Reform in China
Kjeld Erik Brødsgaard

119 Religion and Ecological Sustainability in China
Edited by James Miller, Dan Smyer Yu and Peter van der Veer

120 Comparatizing Taiwan
Edited by Shu-mei Shih and Ping-hui Liao

121 Entertaining the Nation
Chinese television in the twenty-first century
Edited by Ruoyun Bai and Geng Song

122 Local Governance Innovation in China
Experimentation, diffusion, and defiance
Edited by Jessica C. Teets and William Hurst

123 Footbinding and Women's Labor in Sichuan
Hill Gates

124 Incentives for Innovation in China
Building an innovative economy
Xuedong Ding and Jun Li

125 Conflict and Cooperation in Sino-US Relations
Change and continuity, causes and cures
Edited by Jean-Marc F. Blanchard and Simon Shen

126 Chinese Environmental Aesthetics
Wangheng Chen, translated by Feng Su, edited by Gerald Cipriani

127 China's Military Procurement in the Reform Era
The Setting of New Directions
Yoram Evron

128 Forecasting China's Future
Dominance or Collapse?
Roger Irvine

129 Chinese Migration and Economic Relations with Europe
Edited by Marco Sanfilippo and Agnieszka Weinar

Forecasting China's Future

Dominance or collapse?

Roger Irvine

Routledge
Taylor & Francis Group
LONDON AND NEW YORK

First published 2016
by Routledge

2 Park Square, Milton Park, Abingdon, Oxfordshire OX14 4RN
711 Third Avenue, New York, NY 10017

Routledge is an imprint of the Taylor & Francis Group, an informa business

First issued in paperback 2017

British Library Cataloguing in Publication Data
A catalogue record for this book is available from the British Library

Library of Congress Cataloging in Publication Data
Irvine, Roger.
Forecasting China's future : dominance or collapse? / Roger Irvine. –
1st edition.
pages cm. – (Routledge Contemporary China Series; 128)
Includes bibliographical references and index.
1. China–Forecasting. 2. China–Politics and government–2002–
3. China–Economic conditions–2000– 4. China–Environmental
conditions. 5. China–Foreign relations–21st century. 6. Twenty-first
century–Forecasts. I. Title.
DS779.4.I78 2016
303.4951–dc23 2015005414

ISBN: 978-1-138-91675-3 (hbk)
ISBN: 978-0-8153-7115-1 (pbk)

Typeset in Times New Roman
by Wearset Ltd, Boldon, Tyne and Wear

Contents

Acknowledgements

This book is based primarily on doctoral research undertaken at the University of Adelaide in South Australia from 2008 to 2013. I would like particularly to express my sincere gratitude to my principal supervisor Dr Gerry Groot, Senior Lecturer in the Centre for Asian Studies of the University of Adelaide, for his unfailing encouragement, valuable advice, and friendship. He, and my second supervisor Professor Purnendra Jain, provided a stimulating and supportive environment that allowed me to pursue my research independently but with guidance and advice when required. I also thank my third supervisor, Dr Czes Tubilewicz, for his advice during the initial years of my candidature.

My thanks also to: Michael Dutton and Malcolm Cook for supporting my participation in the Griffith-Lowy Joint Project on China's Future; Emeritus Professor Andrew Watson for arranging my research attachment in 2010 to Tsinghua University, Professor Chu Shulong for his supervision during that attachment, and Professor Hu Angang for allowing my participation in his lecture series on China's economy; Matt Durnin for assistance with my article in the journal *China Security*; Michael Wesley and Sam Roggeveen for accepting my series of posts on China's high speed rail in Lowy's *Interpreter* blog; Professors Yang Guang and Wang Xueyu for inviting me to lecture at Shandong University; Flinders University for agreeing to my participation in the 2012 Sustainable Environment conference in Beijing with the Chinese Academy of Social Sciences (CASS); Dr Alex English at Victoria University for including me in the 2012 Low Carbon Cities workshop in Beijing with ERI; Elizabeth Steffey Vinton for inviting me to submit an article in the journal *American Foreign Policy Interests*; the Australian government and University of Adelaide for my Endeavour Award fellowship and Discover China grant; and the Taiwan Ministry of Foreign Affairs for awarding me a Taiwan Fellowship in support of my postdoctoral research.

I would also like to warmly thank all my fellow postgraduates and staff at the Centre for Asian Studies during the period of my candidature for their conversation, friendship, and mutual support. Last but certainly not least, I want to record my gratitude to my family, and especially my wife Sue, for their understanding and patience.

Abbreviations

CASS	Chinese Academy of Social Sciences
CCP	Chinese Communist Party
CGA	Center for Global Affairs
CNP	Comprehensive National Power
DRC	Development Research Center
EIA	Energy Information Administration
ERI	Energy Research Institute
GBN	Global Business Network
GDP	Gross Domestic Product
GFC	Global Financial Crisis
GHG	Greenhouse Gas
IEA	International Energy Agency
IMF	International Monetary Fund
LAM	Linear Ascent Model
LBNL	Lawrence Berkeley National Laboratory
MER	Market Exchange Rates
NIC	National Intelligence Council
NPC	National People's Congress
OECD	Organisation for Economic Cooperation and Development
PLA	People's Liberation Army
PPP	Purchasing Power Parity
SEI	Stockholm Environment Institute
UNDP	United Nations Development Programme
WEF	World Economic Forum

1 Introduction

This book provides a comprehensive survey and critique of expert forecasts about China's future in the post-Mao era. Other researchers have considered aspects of this subject, but this survey is unique in its attempt to review these forecasts systematically. China's future development is likely to have a huge impact on twenty-first century global outcomes. It is therefore surprising that so little attention has been given to comparing and evaluating these expert forecasts and the methods employed. Of course no such survey can be truly exhaustive, given the plethora of forecasts and the fact that new ones emerge almost daily. However, the extensive research on which this book is based gives confidence that the diversity of expert opinion is adequately reflected.

The following chapters give a representative sampling of expert forecasts and judgements about the major issues confronting China as a nation and within four principal domains – political, economic, environmental, and international. For each of these major issues the principal arguments and their main proponents are identified. The findings of this summary record turn out to be of considerable interest. They reveal that on each of the issues there are competing expert views, many of them mutually incompatible. This presents a frustrating dilemma for the large audience of policy-makers, scholars and members of the general public who are anxious to have a better appreciation of major trends in a country whose future impact is likely to be crucial for international peace and prosperity, and for the wellbeing of its own vast population.

Dominance or collapse

While the forecasting efforts of foreign experts – often labelled collectively as China watchers – are both legitimate and defensible, it is immediately striking and clearly problematic that these efforts have yielded a very wide spectrum of opinion about where the country is heading. To be more specific, many forecasts have shown a strong tendency to polarise around either one end of this spectrum or the other. They include firm expectations that China will continue to rise and become dominant in its region or even globally, as well as confident predictions[1] of its steady decline or even imminent collapse. This dominance–collapse dichotomy is very well represented in the literature. The expert debate sometimes has the appearance of a contest between two well-defined opposing camps. Observers might well conclude that there is less unanimity on the question of China's future direction than on most other subjects of major international importance.

One might imagine, in view of the obvious difficulties involved in forecasting most aspects of China's future, that experts would be hesitant about backing one or the other of these polarised viewpoints. On the contrary, the confidence – even hubris – displayed by proponents of these two extremes is notable. The difficulty is that experts on both sides seem able to produce with relative ease a variety of arguments, evidence, and even apparently objective statistics to support their positions. The resultant dichotomy is a serious problem. It undermines the ability of China experts to contribute effectively to two of their most important tasks: namely, elevating public understanding of China's likely future impact, and informing efforts by decision-makers to respond appropriately to China's rise. It obscures vital issues and confounds the formulation of effective responses. It runs the significant risk of leading to sub-optimal and even dangerous outcomes for China's international and domestic relationships.

There is clearly, and understandably, an immense public curiosity about China's future direction. But the majority of interested observers are increasingly bewildered by competing expert arguments and often seem to be swayed one way and then the other, depending on which view is currently most fashionable or voluble. Those who argue China will continue to prosper and become increasingly dominant are often described, more so by their critics than by themselves, as optimists. Those who see the prospect of China's collapse are more clearly worthy of the title of pessimists, and usually accept this categorisation.

The middle ground

This dominance/optimist or collapse/pessimist dichotomy of opinion is well known. It is much less appreciated, however, that there is an equally important middle ground between these extremes. Another large group of experts – including many of the most experienced and respected China watchers – are not at all clear in their minds about whether China will in future be either dominant or prone to collapse. One of the most important outcomes of my research for this study was the discovery that, within the debate about China's overall outlook, and also on particular major issues, a significant body of experts are unconvinced by either of the opposing viewpoints. In fact this cohort of experienced China watchers concede quite readily that China's future prospects are better described as uncertain or unresolved.

Different experts, it appears, respond to uncertainty in quite different ways. Some gravitate with relative ease toward strongly held opinions at the two opposing ends of the available spectrum. They seek to back those opinions with confirmatory evidence and sometimes also with allegedly supportive theory or models. They tend to adhere strongly to their sharply defined opinions, even in the face of emerging contrary evidence. Other experts are drawn instead toward the middle ground because – instinctively or through experience – they are more impressed by the extent of the uncertainties with which they must contend. They are more inclined to be circumspect, more willing to seek out and consider contrary evidence, and more likely to adjust their judgements accordingly. Their ability to see the possibility of more than one outcome often increases their ability to give nuanced advice to decision-makers.

Both types of experts – the bold and the cautious – may suffer from limitations. Those who rush to judgement, hold strong opinions, give preference to personal biases, and are inflexible in the face of new evidence, are clearly fallible. Those who are more tentative and reserve their judgement may sometimes be unable to develop clear conclusions or offer practical advice. My research suggests there is likely to be much advantage in an alternative more-balanced approach that gives objective consideration to the arguments of experts representing several parts of the spectrum, and in regarding over-confident judgements with caution.[2]

The title of this book, while designed to draw attention to the obvious preoccupation of many experts with extreme scenarios, does not imply that either one or the other of these two opposing outcomes is considered likely or inevitable.[3] On the contrary, this study emphasises that there are other approaches that may lead to a more rational response by experts to the presence of major uncertainties. A conspicuous flaw in efforts to forecast China's future has been the failure by most experts to recognise, accept, and embrace the reality and consequences of this overwhelming prevalence of uncertainty about most if not all of the major issues confronting China. The tendency for many expert judgements to polarise around opposing viewpoints is itself a reflection of the fact that the future *is* inherently uncertain. These opposing schools of thought cannot possibly both be correct. The omnipresence of uncertainty is the central problem with which forecasters of China's future must grapple.[4]

Critical uncertainties

These findings led me to investigate whether this uncertainty about most key aspects of China's future could be used as a new starting point, and be harnessed pragmatically as the basis for a new framework for research and analysis. In an attempt to provide a more constructive alternative to the prediction of either China's dominance or its collapse, this book recommends a new approach based on the identification and structured consideration of a hierarchy of *critical uncertainties*. These uncertainties are defined as "critical" because they are identified either nationally or within each domain at a strategic or system level, and because the way in which they unfold will have very high consequences and impacts nationally and internationally.

This approach is applied in Chapters 4 to 7 to the presentation and analysis of expert forecasts within each of the four principal domains. I believe these chapters demonstrate convincingly the merits of an analysis of China's future prospects based on the investigation of such specified critical uncertainties. This approach focuses attention directly on those issues that are of greatest consequence, and on the competing views of relevant experts. It exposes likely shortcomings in the arguments of individual experts by contrasting them with equally plausible counter-arguments. This results in valuable new perspectives. It gives readers the opportunity to consider the strength or weakness of the opposing arguments and perhaps to form their own views.

The domain-level analysis also considers the effectiveness of the forecasting *methods* used by particular experts. This parallel focus on methods is another key aspect of this study. It is based on a thorough prior consideration of the merits and shortcomings of the available and most widely employed methods. Useful, though imperfect, methods are found to be indeed available. Many of them have been employed by China experts, although with mixed success and with significant potential for further improvement.

The final chapter of the book emphasises that for all of the highest-level critical uncertainties discussed there is scope for both positive and negative outcomes. This finding is of great importance. It follows that there is considerable opportunity, through appropriate policies and decisions by key actors, to maximise the potential for positive or desired outcomes rather than the alternative of fatalistically succumbing to equally possible but negative outcomes such as dominance or collapse. Wider recognition by both China and other countries of the realistic prospects for encouraging these positive outcomes is likely to be of great significance for the future management of the array of complex and challenging issues associated with China's development.

Credibility of forecasting

At the outset, any discussion of forecasting or prediction must first confront the fundamental issue of the credibility of all such endeavours. A central question arises in both scholarly analysis and popular perceptions: namely, how can a discipline that seeks to anticipate the necessarily uncertain future have any validity, especially when the majority of these forecasts turn out to be wrong, either in degree or in nature? Forecasting the future of any country – and especially a country like China – is widely accepted to be a notoriously difficult, if not quixotic, enterprise. Retrospective study reveals that, though many attempt it, few have had conspicuous success. Such forecasting is therefore much derided, and many researchers consequently appear to feel more comfortable analysing historical or contemporary developments. They consider it foolhardy to try to peer into the future.

Nevertheless many experts have chosen, perhaps paradoxically, to cast aside this understandable and even commendable caution. In the case of possibly no other country is this more apparent than for China, where there is a prolific and constantly expanding international literature in which determined and well-resourced efforts have been devoted to assessing its future prospects. This is occurring despite the fact that the task is even more difficult in the case of such a large and dynamic nation. Yet the reasons for the growing fascination with China's future are readily apparent. China is the most populous and fastest growing country in the world. Unless current trends change dramatically, it will soon become the world's largest economy. Its growth performance is already a critical factor in the health of the global economy. As the world's largest emitter of greenhouse gases (GHGs) its influence on the world's environment will be equally critical. Many are also concerned that its influence on the international order may be disruptive and that its pursuit of its international interests could lead to military conflict. China's development may therefore have a greater impact on the global future than that of any other country except perhaps the United States. As China's power has grown steadily, an increasing number of experts have therefore taken up the challenge of forecasting its likely direction.

This study has taken the view that resolving the apparent conundrum about the validity of forecasting requires a more pragmatic and even indulgent approach to this inexact art than is often encountered. Against the weight of popular scepticism, it argues that recognition should be given to the following important considerations that support the acceptance of forecasting as a worthy occupation. First, it should be acknowledged that attempts – no matter how imperfect – to anticipate the future as a guide to actions in the present are in any case a universal and inevitable preoccupation for individuals and organisations. Second, planning, policy-making, and decision-making activities that seek to have a positive influence on the direction of future outcomes will in most cases be more effective if based on forecasts than if they proceed without any structured basis. Third, in addition to these persuasive considerations it should be noted that forecasting and planning activities are usually iterative in nature. Forecasts tend to be updated regularly by new studies that take account of new data, trends, and perceptions, and the mistakes of previous forecasts. Inaccurate forecasts may then be rejected or superseded. Fourth, forecasts of national trends increasingly incorporate the possibility of alternative futures. While this reduces forecasting precision, it increases credibility and serves the interests of policy-making by drawing attention to several possible futures that can be either encouraged or avoided. Fifth, the credibility of forecasts can also be increased by giving greater transparency to the assumptions and reasoning on which they are based.

China itself provides compelling evidence for the argument that forecasting is a necessary and legitimate activity. Its future is manifestly too critical a factor – for global prosperity, sustainability, and stability – for estimation of the likely parameters of its future development to be neglected. This study therefore takes the position that not only are expert attempts to forecast China's future going to continue in any case, but such forecasts are also likely to play an increasingly important if not crucial role.

Although some scholars have questioned and even ridiculed attempts to forecast trends in the social sciences, many have been more sympathetic. Probably the most comprehensive overview of the theory and validity of forecasting has been provided by the distinguished US philosopher Nicholas Rescher, who concluded: "Foreknowledge regarding the developments of the future – or even only plausible surmise about it – is of immense practical and theoretical importance to us." Human beings, he noted, have "a vast stake in the future – so much so that the option of abstaining from predictive ventures simply does not exist for us".[5]

China watchers and forecasting

Many China watchers have themselves wrestled with the tension between the difficulty of attempting to see into China's future and the perceived necessity to nevertheless make the attempt. They have included some of the most eminent scholars in the field. Their views are instructive. For example, the difficulties facing China watchers have been well expressed by Australian academic Richard Rigby, who advised that – taking into account China's size and complexity – experts should be "distrustful of any simple descriptors or characterisations". China's unique character meant that expectations it would take a similar path to other countries were "certainly going to be proved wrong".[6] Other experts commented that circumspection was essential because China was so diverse and rapidly changing it almost defied understanding.[7]

Still others pointed to the many forecasting failures of China watchers. David Shambaugh cautioned they should "tread carefully before hazarding predictions" because their efforts were "littered with predictive casualties". In Sinology as in seismology, he concluded, it was "impossible to forecast with any precision".[8] Charles Freeman lamented that predictions about China displayed "a remarkable record of unreliability".[9] Zhang Qingmin at Peking University was similarly concerned, finding that China forecasts were "replete with examples of erroneous predictions and glaring failure to anticipate major events".[10]

Despite this poor track record, there were many China watchers who clearly believed the attempt was worth making. They frequently and emphatically endorsed both the validity and necessity of forecasting. Shaohua Hu, for example, accepted that predictions in the social sciences were risky because future developments were "determined by numerous, complicated, changing, and interacting factors". Nonetheless he believed that identifying potential developments was "both possible and necessary: possible because historical events do not come out of the blue, and necessary because one major purpose of understanding the world is to influence its future".[11] Maria Hsia Chang conceded that, in the case of China, attempts by political scientists to anticipate events were "always hazardous and, at best, tentative". But she believed that China's significance was too great to give up the attempt.[12] Cheng Li agreed, because China's trajectory would have "profound implications not only for the millions of Chinese people but also for the rest of the world".[13]

Other China watchers concurred that, no matter how difficult the exercise, it was still necessary to attempt to forecast its future. Peter Gries noted that there were many difficulties but urged these not be used as "excuses to dismiss the enterprise".[14] Alan Whiting considered that "tentative answers to important questions are more useful than easy answers to trivial questions".[15] An ambitious approach was also favoured by Shambaugh, who regretted a tendency for China studies to focus on "increasingly smaller units of analysis" that failed "to generate broader views or predictions", resulting in a field of study that seemed "to know more and more about less and less".[16]

There appears nevertheless to be widespread agreement that the forecasting track record of China experts is at best mixed and at worst woefully inadequate. These failings will be copiously illustrated in following chapters, but it is not the purpose of this study to denigrate the forecasting errors of China watchers. If one accepts the premise that China's future is a legitimate field of speculation that is too important to ignore, then forecasting stands as a worthy endeavour. The determined efforts of China experts to overcome the difficulties they face and to attempt to provide guidance to those who seek to influence China's future are rather to be commended. It is especially important to acknowledge, along with the failures, the wisdom many experts accumulated about how best to approach this daunting task and the valuable contributions they have made. Their failures, while regretted, should perhaps be viewed philosophically, while their successes deserve to be celebrated.

Methods for forecasting China's future

The failure or success of expert forecasts about China is dependent in part on the effectiveness with which they use the available methods. As noted earlier, this book is based on an extensive review of the literature about the variety of methods that may be relevant for forecasting national futures. This in turn was preceded by an investigation of the contribution made by the emerging discipline of Futures Studies. Although speculation about the future dates from ancient times, formal academic study about predicting the future began in the mid-1950s in Europe, spreading quickly to the United States and other countries.[17] Futurists emphasised that they sought to explore not just one future but to identify "possible, probable and preferable" futures.[18] Others, however, believed accurate prediction of the future was not possible because we live in a complex world where uncertainty is dominant.[19]

From an early stage, futurists devoted attention to the development of forecasting methods. Nicholas Rescher devised a logically structured framework of predictive methods in two broad categories. In the first category were *informal* methods – based on *judgement* or intuition, and in the second category were *formal* methods – based on *inference* or rule-specified modes of reasoning and modelling. But probably the most comprehensive catalogue of methods in the Futures Studies literature was provided by Jerome Glenn and Theodore Gordon who identified twenty-five separate methods.[20] From this extensive menu I have selected a set of eight inter-related key methods or approaches that appeared to me to stand out as the most promising for the analysis of China's future outlook. They include all of the most important methods that have actually been used by China experts. These eight methods are illustrated in Figure 1.1.

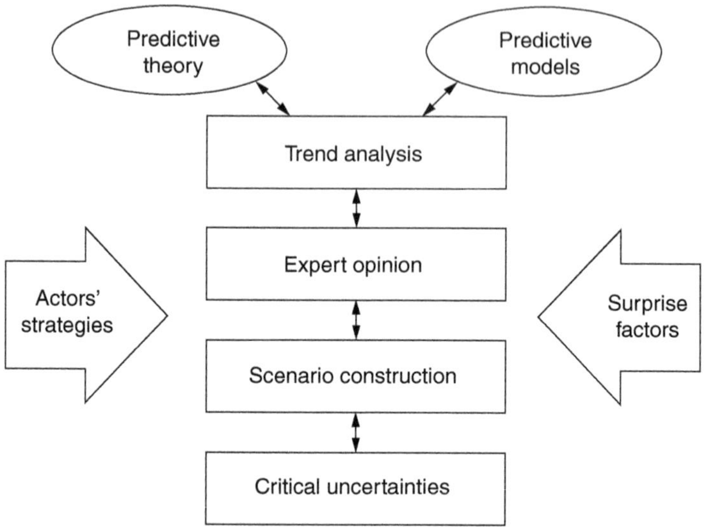

Figure 1.1 Methods for forecasting China's future.

This Figure is illustrative only and does not claim to accurately represent the complex interactions between different methods. Nor do all of the eight methods have equal status, but in combination they offer a viable basis for evaluating expert forecasts about the most significant medium to long-term trends in China.

Trend analysis

The most fundamental forecasting method is the analysis and projection of national trends. It is difficult to imagine any form of forecasting that does not involve the analysis of trends, either qualitatively or quantitatively, or both. Rescher referred to trend analysis as "pattern fitting". He asserted that rational prediction was *only* possible by discerning the emerging patterns and fitting predictions to those patterns. He also emphasised an important precondition for successful trend analysis, namely that relevant information should be available and that it should be timely, accurate, and reliable.[21] Rescher described *trend projection* as the simplest form of trend analysis, but flawed by the fact that prevailing trends could be disrupted by changes in their causal conditions. Trend analysis could also be based on what Rescher called *circumstantial analogy*. First, a country might follow the same course as followed previously by other comparable countries, and, second, a country's future might be strongly influenced, or even determined, by that country's inherited historical or cultural patterns. (Both forms of analogy are familiar in the case of China.) However, such predictive analogies were not reliable because "the extent to which history repeats itself is always limited". Rescher warned that in fact: "No predictive methodology – however carefully contrived it may be – can ever immunize forecasting against the prospect of failure."[22]

Expert opinion

If trend analysis is the fundamental method for forecasting national futures, then the exercise of expert opinion or judgement must surely be the second most essential method – not least because experts are the chief instruments of trend analysis. Especially in the case of China, heavy reliance has been placed on expert opinion. Experts supposedly rely primarily on the analysis of trends, but they are also deemed to acquire expertise from years of research and observation that is believed to empower them to offer special insights and astute judgements about those trends. Their views are therefore widely sought. Paradoxically, however, their judgements are also notoriously diverse and unreliable, and therefore frequently regarded with considerable scepticism.

Philip Tetlock's landmark study on *Expert Political Judgement* provides thorough research and valuable insights concerning the reliability of expert opinion. These insights have informed many of my own views about the fallibility of expert judgement in the case of China. Tetlock's overall conclusions offered little reason to have faith in the reliability of expert analysis. He tracked the predictive accuracy of hundreds of experts across a disparate range of topics and found few signs that their expertise led to forecasts that were any more accurate than random probability. He famously observed that they were unlikely to perform any better than dilettantes or even dart-throwing chimpanzees. Nor did the professional background or reputation of individual experts offer any guide to the accuracy of their forecasts.[23]

Tetlock did make one very notable finding, however, that is of great interest and relevance for this study. He found that *how* experts thought – "their style of reasoning" – did have an influence on accuracy. Borrowing Isaiah Berlin's analogy of hedgehogs and foxes, Tetlock applied these labels to experts and found that

> the intellectually aggressive hedgehogs knew one big thing and sought … to expand the power of that big thing to "cover" new cases; the more eclectic foxes knew many little things and were content to improvise ad hoc solutions to keep pace with a rapidly changing world.

Tetlock found that the predictive abilities of the cautious foxes consistently edged out the opinionated and defensive hedgehogs. Moreover, the foxes "enjoy their most decisive victories in long-term exercises inside their domains of expertise". He observed that "the foxes' self-critical, point-counterpoint style of thinking prevented them from building up the sorts of excessive enthusiasm for their predictions that hedgehogs, especially well-informed ones, displayed for theirs". Foxes were more sensitive to the influence of contradictory forces and "hedged their bets by rarely ruling out anything as 'impossible'". This supports my own conclusion that analysis of China's future is likely to be more fruitful if experts emulate the cautious, empirically based approach of the foxes. But Tetlock conceded there could be some benefits in the "close-minded" approach of the hedgehogs, and found some costs in the "open-minded" approach of the foxes.[24]

Substantial efforts have been made to develop techniques to increase the reliability of expert opinion. One potentially valuable approach was to pool expertise through panels, debates, conferences, and structured exercises such as Delphis. The Delphi survey originated within the RAND Corporation and uses expert panels to address forecasting issues. The technique is designed to remove impediments to expert consensus that arise when experts meet face to face. Anonymity is achieved by consulting experts through individual written surveys. Consensus is encouraged by feeding individual views back to the group and then seeking further rounds of individual responses. The primary strength of Delphis, according to Gordon, is their ability "to explore, coolly and objectively, issues that require judgement". They can assist to crystallise reasons for "dis-sensus", even when they fail to reach consensus. However, they are typically lengthy and time-consuming affairs, and can fail if they do not include sufficient expertise.[25] Armstrong also found that Delphi groups were "substantially more accurate than individual experts and traditional groups".[26] In this study I encountered only one instance where the Delphi technique was employed, and it proved to be quite effective.

Scenario construction

The chief advantage of the scenario construction method is that it acknowledges and embraces the possibility of alternative future outcomes. Given that human actions and unforeseen events will in practice play a major role in deciding these future outcomes, a method that emphasises the exploration of several possible futures has considerable appeal. Scenario construction gives due recognition to the presence of uncertainty. It can compensate for the fact that some widespread assumptions about likely future outcomes may be exaggerated and highlight that other (perhaps more desirable) outcomes are possible or plausible. Scenario construction is therefore potentially very valuable for policy and decision-making. This helps to explain why it is one of the most popular and widely used futures methods, including in the literature about China's future.

A well-known approach to scenario construction and planning is that offered by Peter Schwartz in his *The Art of the Long View*. He suggested using "driving forces" as the building blocks for scenarios. He advocated the following sequence of steps: identify the focal issue or decision; identify the key forces and trends in the environment; rank the driving forces and trends by importance and uncertainty; select the scenario logics; fill out the scenarios; assess the implications; and select the leading indicators and signposts for monitoring purposes.[27] Bernstein *et al.* advocated a similar approach that concentrated on the identification of driving forces, predetermined elements, critical uncertainties and wild cards to develop scenario plot lines.[28] Scenario construction can be difficult to apply successfully, however, and has attracted considerable criticism. In retrospect, many scenarios can appear hopelessly naïve. Armstrong cautioned against using forecasting scenarios because "you are likely to be both wrong and convincing".[29]

Predictive theory

Relatively limited attention has been given to study of the use of predictive theory for forecasting the future. Rescher was one exception. He noted that there were two potential obstacles to reliance on predictive theory – first, the "intervention of chance and chaos" and, second, the intrusion of change, which might "destabilise all attempts at formulating descriptive laws".[30] These considerations are certainly relevant to the of use predictive theory to forecast national futures, in China or elsewhere. Supposedly predictive theories have been used to support the notion that China's political governance will be increasingly problematic if it does not move toward democracy, or that its economy will be increasingly untenable if it does not fully embrace free markets, or that its rise will inevitably lead to conflictual relationships with other great powers.

Some social scientists in the fields of international relations and political science have reflected on the possible predictive potential of their disciplines. Many of their observations are also cautionary. Robert Jervis listed several persuasive reasons why it is difficult to predict the course of international politics – including that actors may alter their behaviour, interaction effects and unintended consequences may intervene, chance and accidents can "send world politics down quite different paths", and the past is not a reliable guide to the future.[31] Nazli Choucri also warned that world affairs are complex and nonlinear. She observed that most forecasters "tend to adopt a position of persistent pessimism, or persistent optimism, or a random mixture of the two, without a solid underlying rationale" and that this amounted to the introduction of "systematic bias". But "the most common error of all", she found, was a parochial view that led forecasters to place undue emphasis on certain variables without taking a wider perspective.[32]

Even Hans Morgenthau – a pioneer of the realist school of international relations – recognised the limitations of predictive theory. He thought the first lesson students of international politics should "learn and never forget" was that the complexities of international affairs "make trustworthy prophecies impossible" and that this was where "the scholar and the charlatan part company". Knowledge of the forces that determined politics among nations was limited, he emphasised, because ambiguous and contradictory tendencies are at play. Which tendency would prevail was "anybody's guess".[33] This advice has sometimes been ignored by realists of a more determinist inclination, as will be shown in Chapter 7. John Gaddis also questioned the predictive value of international relations theory. He found that theorists, despite their awareness of the occurrence of abrupt change, tended to bias their historical and theoretical analyses "too much towards continuity". The "hard" sciences had learned this lesson and had developed "a new understanding of complexity, chaos, and catastrophe", but social scientists had still not embraced uncertainty and unpredictability.[34]

Predictive models

Predictive models have also been used very widely to forecast aspects of national futures, especially in the economic and environmental domains. It is possible in fact that more resources are employed in statistical modelling than in any other national futures method. Such models are supported by enormous databases and by specialist staff in international organisations, governments, think tanks, research institutions, academic bodies, investment advisory organisations, and major businesses. China is no exception.

Much commentary about predictive models has been quite negative, however. Glenn and Gordon surveyed several statistical modelling techniques which, they said, assumed that "information contained in historical data can be extracted, analysed, and reduced to one or more equations that can be used to replicate historical patterns". But modellers often made questionable assumptions, including that "all information needed to forecast desired aspects of the future is contained in selective historical data", that "model structure validly replicates the real life structure of the system", that "the ongoing structure of the system ... will be unchanging", and that deterministic, single-point forecasts were useful.[35]

Rescher noted that predictive models had become "the high fashion of the day" and "the instrumentality of choice in forecasting". In some areas they had been very successful, but he also noted important limitations inherent in their ambitious scope and scale. They incorporated "so vast a series of suppositions and assumptions that the resulting structure, however interesting, has all the solidity of a house of cards erected on a foundation of sand". The more complex and elaborate the model became, and the more parameters it included, "the more room we create for errors and their amplification". Data insufficiency and inability to cope with real-world complexity were major obstacles. Although such models offered users a high level of rational comfort, Rescher thought it was fortunate that informal judgemental processes that dispensed with the need for such detailed theories and models were also available.[36]

Despite such serious charges, predictive modelling continues to be used extensively, especially in the field of economics where it is known as *econometrics*. Econometric forecasts for China have a long history.[37] They have provided some of the most methodologically sophisticated forecasts of China's future available, even though they are subject to all of the potential limitations that apply to other predictive models.

Actors' strategies

An essential but much-neglected consideration in forecasting national futures is the quite likely powerful if not decisive influence of the strategies, actions, reactions, and perceptions of human actors. The quality and competence of a government's own responses to a range of major challenges and the degree to which it is thereby able to maintain popular support are of vital importance to a nation's prospects. Various actors – including governments, leaders, political factions, civil society, and representatives of popular opinion – can often react forcefully to evolving trends in ways that modify or even reverse them. This is especially evident in the case of China, where to date the Chinese Communist Party (CCP) has managed repeatedly to surprise observers with its ability to adapt to changing circumstances, to manage pressing national issues, and to avoid falling victim to problems that many thought would ensure its downfall. Its success is only partial, however, and whether it will be able to continue this delicate balancing act is a major question for the future.

Gordon and Glenn highlighted the importance of studying *actors' strategies.* The future is "never totally predetermined", they maintained, because, however influential past trends may be, actors "possess various degrees of freedom … to arrive at the goals they have set themselves". Attention therefore needed to focus on "those actors who directly or indirectly control the key variables".[38] Gaddis stressed that a major difficulty for social scientists who wished to anticipate the future was that they are dealing with human beings who are capable of "seeing the future take shape" and of devising measures "to hasten, retard, or even reverse trends".[39]

Surprise factors

There is also a growing school of thought in the uncertain twenty-first century that it is necessary to be aware of, and if possible make allowance for, the near inevitability that major unanticipated surprises will also have a potentially huge impact on national futures. This is of course a very difficult area in which to develop a coherent method, given that such surprises are by definition almost impossible to anticipate. Nevertheless there is a strong case for forecasts to acknowledge that surprise-free national futures are unlikely, and to ensure they at least try to factor in the possible impacts of "known unknowns" – such as climate change effects, economic downturns, military conflicts, national disasters, health epidemics, terrorist acts, leadership changes, and disruptive technologies.

Perhaps the best-known recent exponent of the notion that the future is increasingly hostage to disruption by unforeseen surprise factors is Nassim Nicholas Taleb, author of *The Black Swan.*[40] His message for those engaged in forecasting and predicting is that they display both too much confidence in the predictability of events and too little awareness of the inevitable impact of uncertainty. Taleb's figurative black swans have three attributes: they lie outside the realm of normal experience, they carry an extreme impact, and we concoct explanations after-the-fact that make them seem more predictable.[41] Many now believe that the most threatening source of future global surprises may be the unpredictable consequences of global warming and climate change. Growing international concern about these threats has given credibility to the possibility – or even the likelihood – of catastrophic global futures. China, as the world's largest and fastest-growing carbon emitter, seems destined to play a major part in international efforts to limit the most damaging effects of climate change, as will be discussed in Chapter 6.

Critical uncertainties

Unlike the other methods described above, the concept of *critical uncertainties* has not been developed as a clearly articulated method for analysing national futures. It is true that the term is occasionally encountered in the literature on futures methods and also in expert analysis of China's future. It has more often been used in scenario construction to identify major unresolved issues that could lead to quite different scenario outcomes. John Hoffman, for example, recommended that China experts should focus on the "key driving forces", the "biggest uncertainties", and the "key variables". He developed scenarios that attempted to explore "the critical uncertainties facing China".[42] Two reports in the series of *Global Trends* publications produced by the US National Intelligence Council (NIC) were structured to examine "Relative Certainties" and "Key Uncertainties". The most recent report included a new emphasis on "Game-Changers" that were defined as "critical variables whose trajectories are far less certain".[43] However, the concept of critical uncertainties has rarely been elaborated beyond this level.

Although the study of critical uncertainties is not a well-established or widely employed futures method, I have chosen to include it here for two persuasive reasons. First, the concept offers an effective approach for evaluating the range of forecasts presented over time by China experts. Second, it also has clear potential to be used as the basis for a productive new framework for structuring ongoing research and analysis of possible alternative futures for China, and other countries. Such a structure would enhance appreciation of the diversity of expert opinion on major issues and increase awareness of the realistic potential for alternative outcomes.

The study of critical uncertainties, as proposed here, involves researching a set of major and consequential issues about which there is in fact an energetic debate among expert forecasters. The first step in this process is to attempt to identify and define these critical uncertainties. As noted earlier, they are defined as "critical" because they are specified at a strategic or system level and because they will have a very high national impact. A hierarchy of these critical uncertainties can then be used both to analyse the ongoing discourse and to provide a framework for further research. The use of such a structured and iterative process maximises the value of existing research and could perhaps also stimulate more comprehensive and better-coordinated institutional approaches to the study of national futures.

The concept of critical uncertainties captures the intention behind other similar terms – such as "drivers", "major issues", and "key questions" – but emphasises the special need to identify those uncertainties that are *most* critical for national outcomes, to investigate the competing arguments among analysts about these uncertainties, and to remain open-minded about how they may evolve. This approach emphasises the necessity to keep in mind that the issues under investigation *are* uncertainties, even though many analysts choose to under-estimate or even ignore the level of uncertainty that is present. The process that I advocate involves identifying, defining, and periodically revising, a relatively short list of the most significant uncertainties nationally and within each of four principal domains – political, economic, environmental, and international. These critical uncertainties are probably most conveniently expressed as a hierarchy of domain-specific interrogative questions. There will naturally be no one correct or enduring statement of critical uncertainties for any particular country. Rather, they will need to be flexibly redefined and revisited as required. They are also quite likely to reflect the different preoccupations of individual researchers.

I will seek to demonstrate in Chapters 4 to 7, and confirm in Chapter 8, that there is an excellent case for a method based on the consideration of critical uncertainties to be given greater prominence, especially for considering the most critical and determinative national issues. Within each of the four domains it is usually not difficult to identify some central and recurring issues that define much of the ongoing debate. The adoption of such a method would reinforce acceptance that two or more alternative outcomes may potentially eventuate for each of the critical uncertainties, rather than perpetuating the not uncommon situation where individual experts doggedly compete in championing their own particular view. This new approach could also encourage concerted promotion of the most desirable futures, rather than premature acceptance of sub-optimal outcomes.

In Chapters 4 to 7 the identification of a series of critical uncertainties has been used as the principal means to structure an investigation of competing expert views and approaches. In large part, readers are left to judge for themselves the merits of the various assessments. However, it is likely to become apparent that prevailing judgements are usually open to challenge from new research or different viewpoints that can easily be overlooked in a tendency by many experts to rush to judgement. Forecasting in circumstances of continuing uncertainty is clearly the quintessential task facing those China experts who take an interest in discerning its future trends.

Sources

This book is based primarily on extensive research of English-language, publicly available, written sources in which experts have made a deliberate attempt to forecast important aspects of China's future. This is a very large field of literature, amounting in this case to over 1,000 documents. Nevertheless, given the volume and diversity of such information, and with more material constantly emerging, it is inevitable that some important sources will have been overlooked.

There is an unintended prominence of Western and especially US sources in this literature, simply because this happens to be where most of the forecasts have originated.[44] The views of experts in China – including English-language translations and items published by these experts overseas – have, however, been sought out and used where relevant. These Chinese sources are not as numerous or accessible as might be desired, partly because most of these experts have typically been more interested in planning for China's future than in predicting it and partly due to the political constraints under which they operate. There are important exceptions, however. Many senior Chinese researchers appeared able to publish material about important and sensitive issues affecting China's future, especially in the economic, environmental, and international fields. This often occurred in close collaboration with overseas counterparts and international organisations. Émigré Chinese who were born and/or educated in China, as well as other scholars of Chinese ethnicity, have also contributed greatly to the literature about China's future, usually with backgrounds that give them valuable linguistic and cross-cultural advantages. These sources have also been used extensively.[45] Finally, with regard to Chinese sources, it is worth noting Richard Rigby's perceptive observation that there is in contemporary China a vast world of debate "as much within the organs of state and party as outside". He notes that some of this debate is visible, some largely hidden, but it is here that "the vital question of what sort of China we are going to be dealing with in 10, 20, or 30 years from now is being worked out".[46]

My main concern regarding sources is that I am conscious of a burden of responsibility to reflect as accurately as possible the published forecasts of the China experts I have cited. The pragmatic research methodology used in this study – which has been labelled variously as "synthesis", "meta-analysis", or "meta-synthesis"[47] – seeks to identify the relevant publications, to examine the methods used, and to present a narrative or synthesis of their most important findings. My aim is to compare competing views and hold them up for scrutiny. This process involves many pitfalls, especially the danger of subjectivity and lack of precision in summarising expert arguments. Mindful of these potential problems, this study does not underestimate the ambitious scope of its objective and the complexity of its subject matter. The approach taken necessarily requires extensive reference to the writings of individual experts, which is unavoidable if the aim of accurately but succinctly presenting their views is to be achieved. In addition, my analysis goes beyond presentation and comparison of views to deeper analysis of the way in which experts respond to uncertainty and the effectiveness of their forecasting methods. I am mindful that in doing so there may be a risk of over-simplification or even distortion. I have therefore taken particular care to adhere closely to the specific arguments they have employed. To the extent that this may nonetheless have been insufficient in some instances, I can only accept responsibility for such shortcomings and plead the mitigating circumstance of the risks inherent in any major synthesis exercise.

Overview

The following overview describes how subsequent chapters of this study have been structured.

Chapter 2, *Pessimists and optimists*, charts the development of expert judgements about China's national outlook since the beginning of the post-Mao period in the late 1970s. It notes the increasing optimism of analysts during the 1980s, their very negative assessments following the 1989 Tiananmen crisis, and the new expectations that emerged in the 1990s as a consequence of growing appreciation of the reality of China's rise. It highlights the emergence of a marked polarisation of expert opinion between, on the one hand, those who were strongly sceptical about China's prospects and argued the case for pessimism and, on the other, the mostly more cautious optimists who believed it likely that major challenges would be overcome and that a case could be made that China was likely to continue to prosper and expand its influence. It notes that this bifurcation of opinion became a distinctive feature of the expert debate during the 2000s and continues to be important, subsuming a parallel and more radical dialogue between those who saw the prospect of China's collapse and those who thought it was likely to become dominant.

Chapter 3, *Acknowledging uncertainty*, considers the major concerns raised by the evident unreliability of expert opinion but concentrates on examining the extensive middle ground between the pessimists and the optimists. This middle ground is occupied by a significant number of experts who avoided those extremes and were more impressed by, and accepting of, the uncertainty surrounding China's future. The chapter notes a somewhat different approach taken by those experts who focused more narrowly on the question of "What could go wrong?" to upset China's rise. It considers the limited contributions made by expert groups. It also describes the use made by China experts of the scenario construction method and evaluates its effectiveness. It emphasises the value of scenarios in raising the possibility of alternative outcomes, but notes some important limitations. In particular, it comments on the difficulties that arose when experts declined to assign probabilities to their scenarios and the apparently greater success enjoyed by those who took up this challenge.

Chapters 4 to 7, *Political uncertainties*, *Economic uncertainties*, *Environmental uncertainties*, and *International uncertainties*, mark a shift in attention from expert views about China's overall national future toward consideration of expert judgements about critical uncertainties within the four major domains – political, economic, environmental, and international. The argument made here is that a focus on such critical uncertainties offers not only a useful template to examine the ongoing debate among China watchers, but also a valuable research framework for comparing and assessing competing judgements. Each of the chapters concludes with a *Futures* section that comments on the important new perspectives arising from the investigation of critical uncertainties and on the effectiveness and limitations of the other specific forecasting methods and techniques used.

Chapter 4, *Political uncertainties*, observes that the expert debate has focused rather more on the future institutional form of China's political system than on the broader question of the likely effectiveness of political governance. It describes a wide spectrum of expert views, ranging from predictions of short-term or long-term democratic transition to expectations that the authoritarian system will survive and scepticism about the prospects for political liberalisation within that system. It notes that many experts questioned the use by their colleagues of predictive theories to forecast China's political future.

Chapter 5, *Economic uncertainties*, documents the considerable divergence, over a long period, in quantitative forecasts of China's economic growth rate and especially the persistent underestimation of that growth. It describes similar disagreement among experts about when, or even if, China's GDP will overtake the United States. It highlights the extent to which this discussion is handicapped by the regrettable lack of a credible and widely accepted measure for comparing the size of national economies. It also examines the unresolved and highly consequential expert debate about China's prospects for transitioning successfully to a more sustainable long-term economic model.

Chapter 6, *Environmental uncertainties*, considers evolving and contrasting expert assessments about China's ability to manage its still-growing environmental challenges. It considers the vital issue of whether China's future energy consumption and carbon emissions will be contained by an impressive array of counter-measures, or whether these measures will be overwhelmed by continued economic growth. It includes analysis of several major statistical modelling exercises that have projected the likely future growth of China's carbon emissions under different scenario assumptions. It highlights a surprising degree of cautious optimism, based on China's success in moving toward a less carbon-intensive economy and on the possible prospects for further constraining its emissions, but tempered by substantial residual uncertainty and concern about future challenges.

Chapter 7, *International uncertainties*, analyses the continuing vigorous discussion among experts both outside and inside China about whether its international relations are likely to be predominantly competitive or cooperative. It considers growing concerns about China's alleged assertiveness in pursuit of its core national interests, along with evidence that it still places much emphasis on encouraging peaceful and cooperative relationships. It discusses the predictive claims of international relations theories, and numerous objections to those claims by both Chinese and Western scholars. It also emphasises the important influence of actors and perceptions.

Chapter 8, *Embracing uncertainty*, summarises the merits and shortcomings of the methods used by forecasters of China's future. It commends experts for the degree of success they have achieved with these imperfect instruments and notes that many show considerable awareness of the pitfalls they face. In addition to the traditional forecasting methods, however, this concluding chapter strongly commends the likely advantages of increased emphasis on identifying and analysing critical uncertainties in China's future, including greater reliance on systematic, collective and institutional approaches. It discusses both negative and positive potential outcomes in each of the four domains and urges that giving greater attention and support to the achievability of positive outcomes would be a considerably more beneficial approach than continued fixation on, and apparent acceptance of, the negative outcomes envisioned in the dominance–collapse dichotomy.

Notes

1 The terms "forecasting" and "prediction" are often used interchangeably. In this study the term *forecasting* is generally preferred when referring to a variety of approaches for estimating the likelihood of a range of future possibilities, while the term *prediction* is used when specific or single-point forecasts are attempted. The main reason for making this distinction is that forecasting is much more likely to be achievable than prediction. This is so because forecasts are usually conditional and can include a range of possible outcomes and timeframes.

2 As made clear later in this chapter, my views about the reliability of expert opinion have been influenced by the findings of Philip Tetlock in his insightful book, *Expert Political Judgement: How Good Is It? How Can We Know?* (Princeton, NJ: Princeton University Press, 2005).

3 An alternative, but also more opaque, title for this book might have been "Forecasting China's Future: Experts and Uncertainty", which in fact was the title of the PhD thesis on which it is based.

4 The presence of such uncertainty is also an important reason why this book wisely does *not* attempt to develop a new methodology or theory of its own to forecast China's future. Nor does it include any specific new forecasts, preferring instead to focus on analysing the forecasts of the many experts who have boldly undertaken this ambitious venture.

5 Nicholas Rescher, *Predicting the Future: An Introduction to the Theory of Forecasting* (Albany, NY: State University of New York Press, 1998), 1.

6 Richard Rigby, "The Challenge of China", *East Asia Forum* Australian National University (7 February 2010).

7 Michel C. Oksenberg, Michael D. Swaine, and Daniel C. Lynch, *The Chinese Future* (Los Angeles, CA: Pacific Council on International Policy and RAND Center for Asia–Pacific Policy, 1997), 6.

8 David Shambaugh, *China's Communist Party: Atrophy and Adaptation* (Berkeley, CA: University of California, 2008), 39 and 170.

9 Charles W. Freeman, "China in the Times to Come", *The Globalist* (21–24 May 2007).

10 Zhang Qingmin, "Book Review: Jae Ho Chung, Charting China's Future", *China Review International* 14, no. 2 (Fall 2007), 390–393.

11 Shaohua Hu, *Explaining Chinese Democratization* (Westport, CT: Praeger, 2000), 154.

12 Maria Hsia Chang, "China's Future: Regionalism, Federation, or Disintegration", *Studies in Comparative Communism* 25, no. 3 (September 1992), 211.

13 Mercy Kuo, Andrew Marble, David Lampton, Cheng Li, Pieter Bottelier and Fenggang Yang, "China in the Year 2020", *Asia Policy* 4 (July 2007), 29.

14 Peter Hays Gries, "Forecasting US–China Relations, 2015", *Asian Security* 2, no. 2 (2006), 67.

15 Alan S. Whiting and Robert F. Dernberger, *China's Future: Foreign Policy and Economic Development in the Post-Mao Era* (New York: McGraw Hill, 1977), 85.

16 Shambaugh, *China's Communist Party: Atrophy and Adaptation*, 23.

17 The early history of Futures Studies is traced by Wendell Bell, *Foundations of Futures Studies: History, Purposes, Knowledge: Human Science for a New Era, Volume 1* (New Brunswick, NJ: Transaction, 2003), 1–70; Rescher, *Predicting the Future*, 19–27; and Edward Cornish, *Futuring: the Exploration of the Future* (Bethesda, MD: World Future Society, 2005), 170–180. The related discipline of Forecasting is introduced in Spyros G. Makridakis, Steven C. Wheelwright, and Rob J. Hyndman, *Forecasting: Methods and Applications* (New York: Wiley, 1997) and J. Scott Armstrong, ed. *Principles of Forecasting – a Handbook for Researchers and Practitioners* (Philadelphia, PA: Springer, 2001).

18 Bell, *Foundations of Futures Studies*, 81, and Eleanora B. Masini, "Rethinking Futures Studies", in *Rescuing All Our Futures*, ed. Ziauddin Sardar (Westport, CT: Praeger, 1999), 40.
19 Spyros Makridakis and Nassim Taleb, "Living in a World of Low Levels of Predictability", *International Journal of Forecasting* 25 (2009), 840–844.
20 Jerome C. Glenn and Theodore J. Gordon, *Futures Research Methodology v2.0 (CD-ROM)* (American Council for the United Nations University, 2003).
21 Many forecasting efforts, including in the case of China, are significantly undermined from the beginning by insufficient, inaccurate, inconsistent, or misinterpreted empirical data. Given China's size and complexity, it presents an especially difficult challenge for accurate data collection and statistical analysis.
22 Rescher, *Predicting the Future*, 86, 97–99 and 101–104.
23 Tetlock, *Expert Political Judgment*, 20.
24 Ibid., 2, 20–21 and 72–92.
25 Glenn and Gordon, *Futures Research Methodology v2.0 (CD-ROM)*, Chapter 3.
26 Armstrong, *Principles of Forecasting*, 125.
27 Peter Schwartz, *The Art of the Long View: Planning for the Future in an Uncertain World* (St Leonards, NSW: Currency Doubleday, 1996), 100–117 and 241–248.
28 Steven Bernstein, Richard Lebow, Janice Stein, and Steven Weber, "God Gave Physics the Easy Problems: Adapting Social Science to an Unpredictable World", *European Journal of International Relations* 6, no. 1 (2000), 53–59.
29 Armstrong, *Principles of Forecasting*, 517.
30 Rescher, *Predicting the Future*, 106–107.
31 Robert Jervis, "The Future of World Politics: Will It Resemble the Past?", *International Security* 16, no. 3 (Winter 1991/92), 39–43.
32 Nazli Choucri and Thomas W. Robinson, eds. *Forecasting in International Relations: Theory, Methods, Problems, Prospects* (San Francisco: W.H. Freeman, 1978), 72–75.
33 Hans J. Morgenthau, *Politics Among Nations: The Struggle for Power and Peace*, Fourth ed. (New York: Alfred A. Knopf, 1967), 19.
34 John Lewis Gaddis, "International Relations Theory and the End of the Cold War", *International Security* 17, no. 3 (Winter 1992/93), 52–54. See also Gabriel A. Almond and Stephen J. Genco, "Clouds, Clocks, and the Study of Politics", *World Politics* 29, no. 4 (July 1977), and Bernstein *et al.*, "God Gave Physics the Easy Problems."
35 Glenn and Gordon, *Futures Research Methodology*, Chapter 10.
36 Rescher, *Predicting the Future*, 107–110.
37 See Lawrence R. Klein and Shinichi Ichimura, eds. *Econometric Modeling of China.* (London: World Scientific, 2000).
38 Glenn and Gordon, *Futures Research Methodology*, Chapter 7.
39 Gaddis, "International Relations Theory and the End of the Cold War", 55.
40 Nassim Nicholas Taleb, *The Black Swan: The Impact of the Highly Improbable* (London: Random House, 2007).
41 Nassim Nicholas Taleb, "The Black Swan: The Impact of the Highly Improbable", *New York Times* (22 April 2007). The conviction that all swans are white was once regarded as an unassailable scientific fact, only to be overturned by the discovery of black swans by the first European settlers in Western Australia.
42 W. John Hoffmann and Michael J. Enright, eds. *China into the Future: Making Sense of the World's Most Dynamic Economy* (Singapore: John Wiley & Sons, 2008), 16–21.
43 These reports are discussed in Chapter 3. Examples of the use of critical uncertainties in scenario analysis are also noted in Chapter 3.
44 There is also a lesser bias toward Australian sources, simply because they were more likely to come to my attention than other foreign sources. They perhaps have the merit of helping partially to offset US-centric views.

45 This study resulted in increased awareness of the importance of further research regarding the influential but often neglected views of China's own experts, especially in the field of international relations.

46 Rigby, "The Challenge of China".

47 For description and evaluation of these methodologies see Claire Howell Major and Maggi Savin-Baden, *An Introduction to Qualitative Research Synthesis: Managing the Information Explosion in Social Science Research* (London: Routledge, 2010), 1–22; Denis Walsh and Soo Downe, "Meta-synthesis Method for Qualitative Research: A Literature Review", *Journal of Advanced Nursing* 50, no. 2 (2005), 204–211; and Deborah L. Fingeld, "Metasynthesis: The State of the Art – So Far", *Qualitative Health Research* 13, no. 7 (September 2003), 895–897.

Bibliography

Armstrong, J. Scott, ed. *Principles of Forecasting – A Handbook for Researchers and Practitioners*. Philadelphia, PA: Springer, 2001.

Bell, Wendell. *Foundations of Futures Studies: History, Purposes, Knowledge: Human Science for a New Era, Volume 1*. New Brunswick, NJ: Transaction, 2003.

Cornish, Edward. *Futuring: The Exploration of the Future*. Bethesda, MD: World Future Society, 2005.

Glenn, Jerome C., and Theodore J. Gordon. *Futures Research Methodology v2.0 (CD-ROM)*: American Council for the United Nations University, 2003.

Major, Claire Howell, and Maggi Savin-Baden. *An Introduction to Qualitative Research Synthesis: Managing the Information Explosion in Social Science Research*. London: Routledge, 2010.

Makridakis, Spyros G., Steven C. Wheelwright, and Rob J. Hyndman. *Forecasting: Methods and Applications*. New York: Wiley, 1997.

Rescher, Nicholas. *Predicting the Future: An Introduction to the Theory of Forecasting*. Albany, NY: State University of New York Press, 1998.

Taleb, Nassim Nicholas. *The Black Swan: The Impact of the Highly Improbable*. London: Random House, 2007.

Tetlock, Philip E. *Expert Political Judgment: How Good Is It? How Can We Know?* Princeton, NJ: Princeton University Press, 2005.

2 Pessimists and optimists

Before Tiananmen

During the course of the post-Mao era, a distinct tendency arose for forecasts by China watchers to polarise between pessimism and optimism, but this was not yet clearly established in the period preceding and following Mao's death in 1976. Perhaps due to the chaotic influence of the Cultural Revolution, few international observers foresaw the likelihood of China's emergence in the future as a stable and prosperous country. As Martin Jacques later noted, it is difficult to imagine who, at the time of Mao's death, would have predicted that China "stood on the eve of a most remarkable period of economic growth that would entirely transform the face and fortunes of the country".[1]

Nevertheless there were rare occasions when those who dared to make a bold attempt to foresee China's future came strikingly close to success. In despatches from Beijing in mid-1976 (later declassified), Australian Ambassador Stephen Fitzgerald raised the possibility that during the last quarter of the century China's economy might expand in the same way that Japan's economy did in the third quarter, and that this might lead to "annual growth in the vicinity of 10% over a period of 25 years" and a ten-fold increase in GDP by 2000. Few China watchers have predicted so successfully. Fitzgerald also ventured the opinion that the period to the year 2000 would see "the extension of dominant Chinese power and influence throughout the region".[2] This more extravagant prediction was, incidentally, possibly also the first of many instances when a dominance scenario was raised.

Other experts were not quite so positive about China's prospects prior to Mao's departure. Eminent US scholar Doak Barnett, writing in 1974, detected a "pervasive atmosphere of tentativeness and uncertainty". Noting that the past record of foreign specialists in projecting China's future trends had been "a good deal less than infallible", he recommended that all projections must be subject to revision and that any serious student of China must "expect the unexpected".[3] Even so, his prognosis was cautiously optimistic. He assessed, perceptively, that, although China would not become an economic superpower in the foreseeable future, its steady modernisation would encourage political and social stability. It would develop its military power and international influence, but maintain a defensive posture and pursue flexible and pragmatic foreign policies. It would be preoccupied with domestic issues. He was unsure whether to be "relatively sanguine" or "comparatively pessimistic", though he favoured "a cautiously sanguine view".[4] Harry Harding was even more uncertain but erred on the side of pessimism. He foresaw the possibility of "a violent struggle for power, widespread social disorder and even economic collapse".[5] This was one of the first occasions when such a collapse scenario was raised.

After Mao's demise, when Deng Xiaoping's reformers progressively took control of the levers of power and launched China's new era of reform and opening-up in 1978, attention focused mainly on China's chances of achieving a modest stability and a return to basic levels of economic well-being. Nevertheless, from an early stage, some China watchers were encouraged by its new stability and early progress in economic development. Although their views remained cautious, many were increasingly hopeful that further progress would be made. One of the earliest concerted attempts to foreshadow China's likely course was made by Alan Whiting and Robert Dernberger, although they sought to deflate exaggerated expectations about their predictive abilities.[6] Whiting ventured that there was no prospect of China becoming a global economic superpower in the next decade or of it facing "utter collapse". He saw China becoming a middle-range power with influence confined mainly to its immediate environment. Dernberger's forecast was for "stable economic growth with moderate foreign policy behaviour".

More optimistic views were not long in emerging, as China's political direction stabilised and its economic management steadily improved. Professional economists were especially positive. Saburo Okita judged that the leadership's realistic and pragmatic approach would ensure that rapid modernisation would continue, and by the end of the century China's industrial capability would rival other major powers.[7] Herman Kahn felt justified in forecasting average GDP growth of around 7 percent for the next two decades. But he added – correctly as it turned out – that growth in the 1990s might accelerate, and that in any case by the end of the century China would be "a very large and rather modern nation".[8] US economist Dwight Perkins also chided those who assumed China would remain far behind the major powers well into the twenty-first century. He thought it reasonable to foresee China's GDP rising after 1980 to 7 percent or more per year and quadrupling by 2000. This, he predicted rather excessively, could put China's power "nearly on a par" with the United States by 2000 or soon thereafter.[9] Some non-economists began to show comparable levels of confidence. Doak Barnett cited a World Bank report predicting that by the 2000s China would rise from a low-income to a middle-income country.[10] He disputed opinions that China's attempt to combine a command economy and a market system was destined to fail. It might also succeed in creating "a historic first", namely a communist nation with "a liberalized form of authoritarianism".[11]

China's actual growth in the 1980s exceeded the expectations of even these optimistic observers, with GDP expanding by an average of over 9 percent per year. Australian analyst Gary Klintworth was one of the few who anticipated that this level of achievement might continue, predicting growth "well within the range of 6–9 percent per annum" for the next 15 years. He thought China might not be a superpower for a long time, but it would loom very large.[12] He proved to be quite accurate, but what neither he nor other experts anticipated was the massive impact of the Tiananmen crisis later that year and the subsequent sharp, though temporary, decline in China's economic performance and international standing. These events led to a dramatic turnaround in expert opinion, from relative optimism about China's prospects to a virtual orgy of pessimism.

After Tiananmen

The shocking impact of the Tiananmen crisis of June 1989 – in which the CCP used lethal military force to suppress a popular movement seeking accelerated political and other reforms – was a watershed event not only in China's modern history but also in expert opinion. A degree of optimism about China's post-Mao development gave way to a profound scepticism. An emotive mood took hold among China watchers. A volume recording their reactions after the massacre edited by George Hicks noted their "anger, impotence and frustration". For years "a thread of optimism – even of wishful thinking" had run through much Western writing about China. But Tiananmen broke that thread and, Hicks claimed, made it clear that "the dissenting minority" who were sceptical about China's reforms had "the clearer vision".[13]

Hicks noted that other contributors shared, to varying degrees, a deep pessimism about China's future. One of the most outspoken was Jurgen Domes, who asserted confidently that the continuation of the CCP into the next century had become "entirely unlikely", and moreover that to assume otherwise was a delusion reflecting "nothing but incompetence in social science analysis". He thought the future of single-party rule in China was "indeed predictable" and that it pointed "unmistakably toward a collapse of communist rule".[14] Several others expressed equally extravagant opinions, with repeated reference to the prospect of collapse. Ramon Myers thought China's cities could become "a tinderbox ready to ignite in flames", plunging China into anarchy and eventual collapse. The CCP sat on "a political powder keg". Michael Byrnes foresaw "the eventual dissolution of the PRC as a nation state". Simon Leys thought China faced an irrepressible tidal wave that would sweep away communism. The regime's collapse was "ineluctable".[15] Similarly, Harvard scholar Roderick MacFarquhar thought that, as in the Soviet Union, "the rot has gone too far". The CCP would "follow the CPSU into the dustbin of history".[16]

A spectacularly gloomy view of China's prospects was later offered by W. J. F. Jenner, who likewise assessed that China was experiencing "a profound general crisis" that threatened its future as a unitary state.[17] A report to the US Congress noted that it was generally perceived that China's leadership was divided, and that reform had slowed dramatically. But it also commented that there were "numerous heated arguments" among US scholars and bureaucrats about where China was going. Most thought it might continue to "muddle through" or perhaps do even better.[18] However, the State Department's Christopher Clarke thought the post-Deng leadership would be forced to rely on coercion and repression. Angry citizens could seek to overthrow the regime.[19] Jan Prybyla assessed that unless China adopted a market system the prospect of "an economic implosion – a house of cards-like collapse" was highly probable within a few years.[20]

All these forecasts in the years following the Tiananmen crisis were handicapped to a greater or lesser extent by their inability to anticipate the major impact of Deng Xiaoping's decision to throw his influence firmly again behind economic reform and modernisation. Deng, approaching ninety, embarked in early 1992 on a symbolic tour of southern provinces during which he denounced leftist obstruction. This was accompanied by a surge in annual GDP growth to a stratospheric 13 to 14 percent.[21] But despite these stabilising trends, several scholars continued to question the viability of the Chinese state. Maria Hsia Chang predicted that a nominally united China could evolve into "a de facto, if not formal, federation", and that its Marxist–Leninist system was threatened with disintegration.[22] Edward Friedman agreed, claiming most Chinese were worried that "the toppling of the Leninist dictatorship in Beijing would lead to the disintegration of China". The old system was "a potentially malignant cancer" maintained by the "frightened old men" in Beijing. Internal tensions were kept in check only by "widespread fear that hate-filled bloody chaos could engulf the country".[23]

A slightly more moderate view was expressed by David Shambaugh, but he also was inclined on balance to be pessimistic, believing China's economic growth was accompanied by social and political decay that was undermining the CCP's rule. The Chinese economic locomotive was "barrelling down the tracks at dangerously high speeds, risking derailment". A "psychology of greed and ruthless opportunism" was taking hold in an increasingly anarchic society. Shambaugh judged this situation was not "a recipe for its imminent collapse or a democratic-inspired revolution", but the regime's legitimacy was declining daily.[24]

A more distinctly negative view was taken by Richard Hornik, who asked whether China was in fact a "Muddle Kingdom". He recommended that optimists should reconsider their "grandiose projections" that China would soon be the biggest economy in Asia if not the world. Its apparent economic success, visible mainly in its cities, was "ephemeral".[25] Gerald Segal also believed China was "a Middle Kingdom in a muddle". There was "much revolutionary tinder scattered around the country", and if it did collapse "no one should say they did not know why".[26] Segal suspected China might be heading for a chaotic era of informal federalism and increasingly devolved power.[27]

By the mid-1990s, scepticism about China's prospects was still very much in vogue among many analysts. David Goodman and Gerald Segal believed the "succession ding-dong" after Deng's death would be protracted. They described the leadership as "the crumbling remnant of an authoritarian system ruled primarily by personality and smoke and mirrors". Politically, China would probably operate as "a looser collection of increasingly independent units". Further deterioration into economic collapse and political chaos was considered unlikely, but not to be dismissed.[28] Willy Lam likewise saw an equally unpromising future for the CCP, which he likened to Cinderella "way past midnight" and with "advanced-stage Parkinson's disease". The CCP's difficulties might open the door for federalism or for greater democracy.[29] James Miles also saw the likelihood of major trouble ahead, including the possible spilling open of a "Pandora's box of rivalry, hatred, vengefulness, and a myriad of other destructive emotions", leading to a chaotic maelstrom whose effects would be felt worldwide. He cited a US Department of Defense forum in which half the participants allegedly believed there would be a breakup after Deng, with China's foreign policy becoming increasingly unpredictable and threatening.[30]

Another significant example of the unremittingly pessimistic tone of so much expert analysis during this period was Jack Goldstone's article arguing the case for "The Coming Chinese Collapse", again testifying to the powerful appeal of the collapse scenario. With a growing population, faltering agricultural production, worker and peasant discontent, and the state rapidly losing its capacity for effective rule, he assessed that China showed "every sign of a country approaching crisis". He doubted that the collapse of communism could, or should, be averted. Given that the CCP was unwilling to grant the democratic reforms that might recapture popular support, Goldstone concluded: "we can expect a terminal crisis within 10 to 15 years". Observers had been startled by the sudden collapse of the USSR, but there would be "no excuse for making the same mistake with China".[31]

During the late 1990s China's growth rate declined steadily to less than 8 percent and only returned to double-digit figures in 2003. This less-than-spectacular decline nonetheless fuelled continued scepticism, leading Gerald Segal to raise the provocative question of whether China really mattered on an international scale. He thought it was at best "a second-rank middle power" that for much of the past 150 years had "consistently disappointed". China was "a small market" with only 3 percent of world trade, and its gross national product was only 3.5 percent of the world total. Its economy was "effectively in recession". In short China was "a classic case of hope over experience", a country that "has great potential, and always will". It was also "a second-rate military power" accounting for only 4.5 percent of global defence spending. It had few friends and was "a political pariah" whose international influence was "clearly puny".[32]

A similarly critical tone pervaded the thinking of London-based investment banker Callum Henderson. Not long ago, he noted, commentators were describing China as becoming "inevitably and inexorably" the world's largest economy, but it was clearly undergoing a "significant and wrenching" economic slowdown that would be likely to defeat this ambition. He saw the potential for "complete financial and economic collapse". China was "on the brink" of either further economic slowdown or a higher stage of development, but he judged it was certainly not inevitable that it would replace the US as the leading economic superpower. As he put it: "what goes up, can – let us not forget! – come down".[33]

The reality of China's rise

Meanwhile the undeniable reality of China's continued rise during the 1990s led in due course to the re-emergence of a qualified optimism, although with growing external apprehension about the implications of its growing power. Not all experts were convinced that communist leadership in China was doomed or that it would be beset with regional disunity, economic decline, and a growing risk of comprehensive collapse. More positive views that took into account the underlying resiliency of the Chinese state were not long in reasserting themselves. Soon after Tiananmen, David Goodman ventured that "remarkably" the situation was on balance "not irretrievable" for the CCP.[34] Harlan Jencks conceded it was probably premature to write off the regime.[35]

David Lampton also perceived China to be less fragile than most other scholars assumed. While acknowledging China's grave problems, he warned against thinking it was "a political and social volcano waiting to erupt" or that it was in a pre-revolutionary state. He argued that the Chinese people had benefited greatly from the reform program and did not want to put this at risk. No other institutions offered an alternative to the CCP.[36] Another nuanced view came from Harry Harding. He accepted that China was "riddled by uncertainties". It was therefore not surprising that post-Tiananmen crisis predictions had been so dire, leading some to conclude the government might collapse "within a few weeks or months". He pointed out, however, that extreme predictions had not been validated, suggesting internal contradictions might not have been "as antagonistic as the pessimists believed". He assessed that there were "powerful integrative forces" that would hold China together.[37]

A notably upbeat prognosis was ventured by *New York Times* journalist Nicholas Kristof, exemplifying a more positive view among many analysts as the 1990s progressed. He was one of the earliest Western commentators to use the phrase "the rise of China", stating that – if it continued – it might be the most important world trend for the next century. China might by 2050 displace the US as the world's largest economy. He injected a note of caution, however, warning of the "tectonic strains" it would face and observing that: "All kinds of things can go wrong, and economic trajectories are almost impossible to predict." Nevertheless, he thought it was at least a realistic possibility that China could sustain its boom for several decades, and that it would be "foolish – and perhaps dangerous – to neglect this likelihood". Some economists believed that under favourable conditions China could maintain growth rates of 7 to 8 percent for "another couple of decades", causing gross domestic product (GDP) to quintuple every 21 years. He argued the West should no longer look at China's "backwardness" as inevitable. It was "on a road that could restore it to its original greatness". His greatest concern was about environmental, energy and climate change issues.[38]

Jack Goldstone's prediction of China's imminent collapse spurred Yasheng Huang to publish an article titled "Why China will not Collapse". He observed that positive projections were usually "drowned out by the growing chorus of doom among China pundits". However, the doomsters were "long on hype and short on facts". He saw no evidence to support cataclysmic predictions. Questionable assumptions had been used to reach sweeping conclusions. Other considerations supported "a relatively optimistic scenario". Huang called for a "more balanced and level-headed" assessment in view of the gap between conventional wisdom and hard evidence.[39]

A thoughtful and comprehensive analysis by Daniel Burstein and Arne de Keijzer also challenged negative views about China. They noted that the pendulum of American thinking had swung "from optimism to pessimism, from hope to fear, from opportunity to threat, from the coming China boom to the coming conflict with China". This swing, they believed, was promoted by those who wanted to portray China as "a new evil empire and a new national enemy" and as "fascistic, expansionist, hegemonic, and inherently adversarial". They sought to offset this alarmism. Their prognosis for China bore in many respects a closer resemblance to subsequent developments. They assessed that China would prosper economically and become influential internationally. It would become in time "the world's largest single nation-state economy". It would challenge assumptions that there was "a high correlation between democracy and economic success". They did not doubt China would be a nationalistic power, but thought that this would not necessarily lead to "overall antagonism or constant friction" with other countries.[40]

Burstein and de Keijzer concluded their assessment with a brave attempt to "fast-forward" to 2024. They foresaw a China that had stayed the course on economic reform and with GDP at near parity with the United States. It would have evolved a "new political–economic paradigm" quite different from the West. China and the US would be "the two leading superpowers" with bilateral relations much improved due to greater liberalisation in China and reduced US concern about it as a strategic threat. In the meantime practically every "dire scenario" ever predicted by Western pundits would have come to pass but "less apocalyptically than was imagined". The CCP in 2024 had held China together, muddling through on most issues. Neither the Hawks nor the Pollyannas had been vindicated.[41]

It is apparent from this summary that by the late 1990s a clear division had begun to form between two camps of experts who seemed to be locked in a contest between those who perceived China's future as problematic and threatening and those who were more inclined to believe it could manage its modernisation process successfully. Subsequently several major contenders emerged in this increasingly central and divisive debate. They presented new or enhanced arguments in favour of their either pessimistic or optimistic views. Their primary weapon was the persuasiveness of their expert judgement, but they also employed other approaches, especially concepts drawn from predictive theory or from ideology. This bifurcation of expert opinion was a notable feature of the

2000s and continues to be an important influence on perceptions of China's future. It subsumed a parallel but more extremist dialogue between those who argued that China faced collapse and those who believed it could achieve dominance.

The case for pessimism

Increasing confidence about China's prospects among some foreign analysts was soon challenged once more at the turn of the century by a vigorous reassertion of the case for pessimism. This school of thought, championed repeatedly and often fervently, has proved to be enduring. A resurgent scepticism attracted adherents with a variety of different viewpoints. Some were pessimistic about China's governance; others were concerned about its economic development, or its environmental challenges, or its social inequalities, or its human rights performance; others were worried about the implications of its international ambitions and its growing military capability. These experts continued to emphasise the possibility – or even the likelihood – that China's political and economic system would collapse spectacularly in spite of its past progress. Some perceived that such an outcome was inevitable and even appeared to relish the prospect of a catastrophic future for China, especially if this cleared the way for political and economic reforms that they deemed to be both essential and desirable.

While many were willing to consider the possibility, and even the probability, that China would collapse under the weight of its numerous challenges and contradictions, it was left to one author in particular to carry the collapse argument to its logical extreme. In 2001 Gordon Chang published his *The Coming Collapse of China*, which has become, at least among English speakers, one of the best-known works about China's possible future. In a sense, there was little that was new or original in Chang's analysis.[42] Several previous authors had explored the collapse thesis, but what distinguished his book was the inevitability and certitude that he attached to it. Few other authors have displayed a similar level of total confidence in their ability to predict China's future. Moreover, Chang's conclusions were little short of apocalyptic. Apart from his very explicit title, his chapter headings were indicative of his clear-cut position – for example, *The Discontent of the People is Explosive, Chinese Banks will Fail, China's Economy Stagnates, WTO Accession will Trigger Collapse, How the State will Fall, The State Begins to Disintegrate.*[43]

Chang claimed that foreigners were so fascinated and dazzled by China's potential that they did not see the facts, namely that China was a "paper dragon" and symptoms of decay were everywhere. It was "in long-term decline and even on the verge of collapse"; boom was turning to bust; the regime was "fundamentally unpopular". His analysis was distinctive in three respects. First, on particular issues the great majority of his judgements were uniformly and uncompromisingly negative. Second, his arguments were distinguished by the frequency with which he resorted to bald assertions – for example: "As time passes, the underlying problems fester. Economic dislocations become social ones, with dark political overtones. At some point there will be no solution. Then the economy and the government will collapse." Third, Chang achieved notoriety for the fact that he (like Goldstone) was willing to attach a specific timeframe to China's collapse. He asserted that: "We are but a few years, perhaps five, from that time."[44]

It is also notable that Chang specifically and categorically excluded the possibility that the state as a key actor might be able to retake the initiative and avert disaster. This was indeed a critical element in his argument, which was that collapse was assured because the CCP was both unwilling and unable to change its authoritarian system. The Party's rule had become "corrupt, repressive and brutal". Even reform in non-political areas was "glacial", like "molasses". The "law of gravity" therefore could not be denied and at some point "the centrally directed system has to fail". Another argument employed by Chang was that scholars tended to treat and downplay China's challenges one by one; but the point, he argued, was that "China faces many challenges all at once, not one challenge at a time". If all these challenges were added together "we can see why the regime will fail". He also argued that even if China could solve its problems in "say, thirty years" that was irrelevant because it did not have the luxury of waiting that long. Its people were "hungry, angry, and worst of all, desperate" and would not tolerate such delay. He claimed that already somewhere in China was the person who would "end the Chinese state as it now exists", because China was like "a lake of gasoline" and that person would only need to "throw a match".[45]

The concluding chapters of Chang's book took on a visionary quality. He declaimed:

> the people are in motion now and one day they will want to govern themselves. That is their destiny.... Sometimes democracy needs a strongman, and at a crucial moment he or she will appear.... The crowd, a million people behind a single man or woman, will cross Changan Avenue and descend upon Zhongnanhai.... It happened in Berlin and Bucharest and it will soon occur in Beijing: the people will take back their government.... And when the world witnesses police clubbing demonstrators and blood in the streets, the reaction will be immediate and emotional.... Across China activists will decide that this is their moment in history.[46]

Not surprisingly, Chang's sweeping conclusions proved controversial. The journal *China Brief* offered a debate on the issues, in which Robert Sutter argued that the Chinese regime would endure. It would be resilient enough to deal with anticipated challenges. It was unlikely to be either "lax or maladroit". It would be aided by popular aversion to chaos and a belief that there was "no viable alternative" to CCP rule.[47] June Dreyer, however, commended Chang for questioning the conventional wisdom and raising the possibility that the People's Republic of China might fall. She believed "it could happen".[48] Chang exercised a right of reply in which he castigated those who did not adopt a similar boldness. He was not surprised by the consensus that the CCP would survive because: "No one, we know, has ever been fired for extrapolating." But when societies began to crumble "timid predictions" would not suffice. He asserted that "the inevitable" would soon occur in China, with the next five years being "the critical period". The Party would "just stiffen and then collapse". He assessed that "all the evidence" pointed to this and that "if anything was inevitable" it was the end of the PRC. He somewhat expanded his earlier timeframe for China's collapse with a clarification that this would happen "within a decade and maybe within five years".[49] In a following article he claimed there had been a stunning reversal of perceptions and that the optimists were "in retreat". "Rosy predictions" were "falling by the wayside" and China was seen as "a failing giant". The "truth", he declared, was that it was "failing economically" and "starting to fall apart". The century that some claimed belonged to China would last "just a decade at most".[50]

By the end of 2003, with more evidence of China's strong economic performance and apparent stability, a different tone seemed evident in Chang's analysis. Uncharacteristically, he was struck by difficult epistemological issues: "As we learn more about the People's Republic, we understand less." There was not one China, but many. Its size and diversity made it "hard to figure out". Observers could "view a sliver and know that portion well" but it was difficult to understand the whole. He also appeared more willing to acknowledge important aspects of China's success. It was apparently a flourishing economy despite its authoritarian political system. It would profoundly affect the future and was "becoming a geopolitical force". The Beijing model might be copied elsewhere, with the developing world taking on "a decidedly China-centric look". But to him this raised another spectre. Even if China was an economic success, he feared it would become an international threat – in fact "we may be creating another Soviet Union, one more dangerous than the first". It could be "a large and economically powerful dictatorial state ... a foe that we're not able to tame".[51]

Soon after, however, Chang returned to his earlier arguments, and added another – that because the velocity of China's economy was now so much faster it had "even less control over what is happening". The state was "courting disaster" due to its unsustainable trajectory. He noted that analysts had a terrible record in predicting turning points in human history and asked: "What are we missing now?" He suggested there might be guerrillas who were "gathering out of sight in the basement" including perhaps hundreds of "secret societies, political parties and revolutionary armies" who wanted to bring down the CCP.[52] In 2006 Chang, still unconvinced by China's uninterrupted progress, wrote that his book had predicted the fall of the CCP "by the end of this decade", and hence it had reached "the halfway point". He conceded China did not look set to fall, but maintained it was "a fragile nation". Protests were becoming larger, more violent, more volatile and more widely based. He still insisted the CCP was incapable of reinvigorating itself. A sharp economic downturn would place it in jeopardy. Popular unrest had the "potential" to cause government collapse.[53]

Chang has continued to defend his position. In a late 2011 article, he admitted that his prediction that the CCP would fall that year was wrong. But, he said, "I'm only off by a year ... don't think I'm taking my prediction back". Deteriorating economic conditions and mistaken policies meant China's "sweet spot" would soon be over. China was "too dynamic and volatile" for the CCP to remain in power. Until recently everything was going well but: "Now, nothing is." He concluded with a rhetorical flourish: "So, yes, my prediction was wrong. Instead of 2011, the mighty Communist Party of China will fall in 2012. Bet on it."[54]

While Chang was not without supporters, few others went so far. But one such was He Qinglian, a former journalist in China, who believed the main reason China had remained stable was due to the state's resort to political violence. The regime concentrated on "pulling up by the roots" all potential sources of instability and on being able to "strangle at birth" any groups that might offer organised resistance. In He's view Chinese society resembled "a volcano on the verge of a major eruption" with "underground fires, smouldering dangerously just below the surface, that could erupt and rage out of control in any place and at any time". The day would come when the CCP would be "consumed in the conflagration".[55] Meanwhile, Venezuelan writer Moises Naim also attached a ten-year timeframe to China's impending crisis. He calculated that China would "almost certainly have an accident within 10 years" that would slow its growth and destabilise its politics.[56]

More significantly, US academic Minxin Pei came to outrival Chang as probably the leading, most articulate, and most persuasive proponent of the pessimistic view. He was one of the first to elaborate the argument that a developing country like China could not modernise successfully without political reform. He drew a parallel between China's situation and the fall of the Suharto regime in Indonesia in 1998. Acknowledging that the notion China could go the same way might seem preposterous, he nonetheless argued that its political system was contending with almost identical weaknesses. A similar crisis might strike China if it failed within the next decade to introduce liberal and democratic reforms. He characterised China as a corrupt and predatory state. In a best case situation it might "muddle through for another decade or two without cataclysmic upheavals", but in the worst case the political system could become so brittle that a leadership split, an economic crisis, or an external shock "could set off a chain reaction leading to a Suharto-type collapse".[57]

Pei's analysis was more reasoned, plausible, and persuasive than Chang's, and his conclusions more qualified, but he was still determinedly pessimistic. He assessed that in the early-2000s China faced a crisis of governance. Its apparent dynamism and promise caused multinational corporations to "salivate", but he argued that "beneath this giddy image of progress and prosperity" was a different reality. Pei and like-minded analysts saw it as essential that China's transition to modernity must involve "not simply a deepening of market liberalization but also the implementation of political reforms that could endanger the CCP's monopoly on power". Pei thought the lesson China had drawn from the collapse of the Soviet Union was that a communist regime could not survive without prioritising economic growth, but it mistakenly believed it could leave the political system untouched. Pragmatists might consider the requirements for rule of law, a market economy, and an open society to be "inconsequential cognitive nuisances" but he believed that they were real. He believed that the CCP exhibited the same pathologies as Brezhnev's Soviet Union and Suharto's Indonesia – "pervasive corruption, a collusive local officialdom, elite cynicism, and mass disenchantment".[58]

In Pei's assessment, the CCP was trapped in "a hopeless dilemma", which was that any reforms that might alleviate social tensions could also trigger a revolution. The situation was "not yet life-threatening", but he predicted China's governance crisis was "likely to continue to grow". This might not cause imminent regime collapse, but accumulating strains could lead to long-term economic stagnation. Hence he advised that the world needed to re-examine its "long-cherished views" about China because they might be "rooted in little more than wishful thinking".[59] Pei argued further that China's economic success had been achieved not so much *despite* its "rotten" political system as *because* it was so rotten. Corrupt CCP officials had "effectively privatised the power of the state".[60]

Pei expanded on his earlier views in his 2006 book *China's Trapped Transition*. He contended that China risked "getting trapped in a 'partial reform equilibrium', where partially reformed economic and political institutions support a hybrid neo-authoritarian order that caters mostly to the needs of a small ruling elite". He believed the future accumulation of governance deficits could eventually lead to "a mass popular political mobilization" that would most probably "precipitate a quick regime collapse". Despite China's awe-inspiring progress a "set of self-destructive dynamics" was making the CCP increasingly anachronistic. He rejected the idea of an "efficacious neoauthoritarian developmental state". He advised the international community that instead of viewing China as "the new superpower of the twenty-first century", it might want to see it as "an underperforming giant that has failed to seize an historic opportunity for making a fundamental break with its authoritarian past and paid a heavy price for it". He viewed China's future through "dark lenses".[61]

In later articles, Pei asserted that the "only thing rising faster than China is the hype about China". He admitted its growth record had "proved pessimists wrong and optimists not optimistic enough", but urged caution "before we all start learning Chinese and marvelling at the accomplishments of the Chinese Communist Party". He argued that the "often breathless conventional wisdom" overlooked major flaws that made many predictions about China's trajectory "misleading, if not downright hazardous". Behind the glowing headlines were fundamental frailties. In short: "China may be rising, but no one really knows whether it can fly."[62] He warned that although its ascendance seemed "unstoppable" its "irredeemably rapacious and self-serving" ruling elites would be likely to be overthrown "when a major crisis hits".[63] He acknowledged that China was "a premier world economic powerhouse", but thought it too soon to regard it as the next superpower, because the likelihood of a significant economic slowdown in the next two decades was "real and even substantial". Internal decay threatened a regime change that would be "disruptive, even cataclysmic".[64]

In early 2009 in the midst of the global financial crisis (GFC), Pei criticised forecasters who thought that, since China had maintained thirty years of rapid growth, this could continue for a similar period. He insisted that it was "time to rethink complacent assumptions" that stemmed from "myopia ... and, worse, a dismissive attitude to warning signs of trouble". He criticised economists for "intellectual tunnel-vision" and for ignoring "elephant-in-the-room risk factors" such as environmental degradation, socioeconomic inequality, and corruption. China's rulers were "sitting on a ticking time bomb".[65] Pei has continued to be pessimistic, suggesting in a 2012 article that it had become "an ingrained habit" to discount the probability of regime change in "seemingly invulnerable autocracies". An alternative perspective was needed that assumed "declining Chinese strength and rising probability of an unexpected democratic transition in the coming two decades". He accepted it was "premature" to write off the CCP's adaptive capacity, but equally its demise could not be ruled out. He saw in China's continuing difficulties "invaluable clues to such a highly probable seismic shift".[66]

Many of the arguments expounded by Minxin Pei were supported, though in a more rigid form, by British author Will Hutton. He asserted that if the next century was going to be Chinese it would only be so because China had embraced the economic and political pluralism of the West in the form of democracy and "genuine capitalism". This would involve adopting what he called "Enlightenment Institutions" such as rule of law, independence of the judiciary, freedom of the press, independent universities, and representative and accountable government – which Hutton considered "indispensable components of any well-functioning economy and society". By contrast, the Chinese economic model – which he described as a halfway house in which the CCP tried to retain political control while allowing "only half a market economy" – was unsustainable. It was structurally unstable and must be strengthened by greater economic pluralism, which would in turn challenge the CCP's pre-eminence.[67]

Like others, Hutton catalogued a long list of governance shortcomings in China and argued that if its growth ever slowed it would be "in mortal trouble". Unless its economic model changed it would be unable to withstand the strain of tripling its economy over the next fifteen years, because the "law of large numbers" would kick in. A tipping point was approaching where, without "self-sustaining entrepreneurial maturity", there would be "a sharp and destabilising deceleration" in China's growth. Hutton concluded that China was caught in a gigantic dilemma. Simple extrapolations of its growth, and predictions that it would become "a one-party economic colossus", were alarmist and naïve. He foresaw a China that was "riven with contradictions and full of potential menace" in a period when it was "coming to terms with the bankruptcy of communism and the logical impossibility of continuing to grow as it has".[68]

Another writer who was equally if not more critical of China's governance and sceptical about its prospects was Australian analyst John Lee. He sought to make "the case for pessimism" by propounding similar arguments to those made previously by Chang, Pei, and Hutton – namely, that significant political reform was unlikely, that market socialism was not a substitute for market capitalism, and that China therefore could not sustain its economic and peaceful rise. Its trajectory, he believed, was "precarious rather than promising". He agreed with other critics that much of China's success was illusory and that necessary reforms were not to be expected because they would threaten the Party's position. He maintained that China's market socialism was "a predatory, dysfunctional and grossly inefficient system that is enormously wasteful and unsustainable". He considered that the CCP was losing its appeal, its organisation was fragmenting, and it was seen as corrupt, self-serving, ineffectual, and irrelevant. The regime faced "deep rivers of discontent" and was "trapped by its own contradictions". It was becoming increasingly desperate and was rapidly losing credibility, fuelling chauvinistic nationalism, and risking crisis over unresolved territorial claims. Economically, "even a slight tremor in the US economy would be devastating". In short, the country was "in a profound mess" and all the ingredients were present for "a potential social or political meltdown".[69]

In a frontal assault on China optimists, Lee lamented that the policies of other countries had become "overwhelmingly predicated on the 'inevitable rise of China', as if there were no other possibility". Most experts, he alleged, were dazzled by modern China, and the presumption that this would be the Chinese century was "undoubtedly gaining momentum". While a handful of other experts were doubtful about China's prospects, Lee claimed the overwhelming consensus in the US, Europe and Australia was "one of general optimism". He believed these experts suffered from "a well-intentioned, but self-induced 'China blindness'". He conceded that in some instances the case for optimism was "well thought out and formidable", but believed that even the most sensible optimists had failed to present compelling evidence that the Chinese authoritarian model could overcome its "profound and mounting contradictions, tensions and dysfunctions". The evidence for pessimism, he believed, was "the more compelling if not yet irresistible conclusion".[70] Lee claimed that many analysts tended to simply produce a "laundry list" of China's successes and shortcomings, and assessed its prospects optimistically or pessimistically depending on which "side of the ledger" they saw as more compelling. But he emphasised the extreme seriousness of China's problems, claiming that the CCP's policies had created "a massive underclass of up to a billion people", with 10–20 percent unemployment in urban areas and 20–40 percent in rural areas.[71]

Somewhat inconsistently, given his argument that China was destined to fail, Lee appeared nonetheless to see it as a threat to Western values and institutions. He quoted Milton Friedman's view that China was "heading for a clash, since economic freedom and political collectivism are not compatible", but he saw Beijing's "stubborn authoritarianism" as "pernicious" and an enormous danger to Western liberal democratisation, because China's success suggested there was "no inextricable link" between political liberalisation and successful transition to a free market. Some even saw authoritarianism as advantageous. This might discourage others from aggressively advocating liberal and democratic agendas. For Lee, the China model was "flawed, unsustainable, dangerously unstable" and its authoritarian logic was "contradictory and duplicitous", but because of its potential appeal to others it presented a moral challenge that was "far more subtle, oblique and sophisticated" than that previously posed by the Soviet Union.[72]

Chang, Pei, Hutton, and Lee all expounded, although with varied interpretations and probabilities, a common set of arguments based on predictive theory which maintained that, without movement toward greater political liberalisation and marketisation, China's rise could or would be undermined. This created an irreducible dilemma because allegedly the CCP knew that the introduction of such reforms would threaten its control. The weakness of these arguments was that China's rise nevertheless continued, and the longer the CCP seemed able to lead this continuing advancement the less credible the pessimists' views appeared. They could, however, always fall back on the claim that the Party was only postponing the inevitable. They could also point to undeniably serious challenges and potentially destabilising factors that were either worsening or were at best contained.

Enthusiasm about China's future prospects was also challenged by Australian expert Paul Monk. He noted a widespread view that China was destined to become an economic superpower, a strategic rival to the United States, and the dominant power in the twenty-first century. This new evaluation of China's potential had occurred "with a rapidity quite as remarkable as the growth of its GDP" which, however, in 2004 was less than 12 percent of US GDP. Monk believed the conventional wisdom was based on "little more than linear extrapolations, decades into the future, of raw and unexamined recent growth rates". It assumed "benign outcomes across a large number of variables crucial to China's possible futures, without critically examining any of them". He referred to this "simplistic and overawed linear way of thinking" as "the Linear Ascent Model (LAM)". He dated the "lurch into LAM-think" to the early-1990s, when the use of purchasing power measures of GDP led to the sudden appearance of an "image of a Chinese economy of massive proportions". He believed such extrapolations "surely belong to the realm of statistical fantasy", and questioned if anyone could seriously believe China's economy would expand to 50 times between 1981 and 2030.[73] He thought it would be "irresponsible to rule out the possibility that China will fail to negotiate some critical thresholds in the early decades of the twenty-first century".[74]

Monk thought "tough questions" needed to be asked of those who conjured up extravagant pictures of China's future. He suggested analysts should balance their "lurid imaginings" against China's complex realities. These realities included its unreformed political system, foreign domination of its export industries, debt, and structural dysfunction – as well as increasingly grave demographic, social and environmental problems. He commended other sceptical analysts who had supplied the appropriate antidote for the optimistic "stuff and nonsense" and "hype and lashings of snake oil" about China's economic prospects. Its economic pressures were "volcanic in their intensity" and could "retard and even derail its growth". He urged forecasters, therefore, to exercise caution and to develop "more serious, multi-dimensional assessments" of China's futures.[75]

Several other authors were also notably sceptical about China's prospects. These included Joshua Kurlantzick, who thought China's economic miracle was "largely a house of cards" and that this façade "probably will crack". Its growth rates were "vastly overstated, the result of cooked books and massive deficit spending".[76] Economics writer Joe Studwell, in his 2003 book *The China Dream*, predicted that at some point in the next five years China would be likely to face a financial and banking crisis. Inefficiencies and corruption were "bleeding the country dry" and signs of China's political and bureaucratic malaise were "inescapable".[77] Scepticism reached new heights with two best-sellers published by Peter Navarro titled *The Coming China Wars* and *Death by China*. Navarro believed China was on "a collision course" with the rest of the world and its future was "at best, highly uncertain". Its "cowboy capitalism" and "amoral foreign policies" were triggering "a whole range of economic, financial, environmental, political, and military tsunamis that threaten to engulf us".[78]

The 2000s also saw some comparatively rare examples of national forecasting by observers within China, although most were more inclined to play down rather than to exaggerate China's future success. These rare exceptions included a paper by three Chinese scholars in 2002 which stated that growing social inequality and unrest in China could lead to violence and the reversal of reform. Instability lay "just beneath the surface of Chinese society" and could be triggered by some unforeseen crisis, such as economic turbulence or adverse external events. This, they ventured, could create a situation similar to that which led to the collapse of the Suharto government.[79] Another unusual instance where a Chinese scholar offered candid and quite negative views was a paper by Nankai University's Zhang Ruizhuang. He advised that those who were "overzealous" about China's rise should heed a Chinese saying that: "Any achievement divided by 1.3 billion is nothing; any problem multiplied by 1.3 billion is overwhelming." He observed that China's growth had been quantitative rather than qualitative, disguising deep structural problems that might be "fatal to the sustainability of China's economic growth". Although average personal incomes had increased twelve times, many had been "left out" and were increasingly resentful. China was "sitting on a powder keg waiting to explode". Externally, its foreign policy had "no principle, no vision, no cause, and no friends". Zhang explained that his assessment was deliberately negative and one-sided, in order to challenge the over-optimism of those who saw only China's sensational achievements.[80]

The case for optimism

Many experts were obviously quite comfortable about forecasting a pessimistic outlook for China. Nevertheless the competing optimistic school of thought survived very much intact. Although optimists often appeared reluctant to be labelled as such – and in fact often mixed their optimism with caution and with professions of uncertainty – they tended at times to be equally as determined as their opponents to win over an increasingly bemused audience. This was perhaps understandable in view of the impressive and abundant evidence of China's continued rapid economic growth and apparent political stability.

As the new century progressed, many experts – especially investment advisors – became increasingly impressed with China's economic prospects. Deutsche Bank concluded that GDP growth of 7 percent or more over the next one or even two decades was achievable if reforms continued.[81] Morgan Stanley believed China could maintain a growth rate of 8 percent or more during the 2000s, propelled by the "single-minded determination" of China's politicians to "make things happen".[82] A much-quoted 2003 Goldman Sachs report marked a new level of optimism. It projected that China's economy, in dollar terms, could be larger than the United States by 2041, and larger than any other economy by 2016.[83] In the following year, however, an update projected it could overtake the US in the mid-2030s.[84]

By the mid-2000s, several experts were making increasingly upbeat forecasts about China's glowing prospects. Oded Shenkar believed many aspects of its future potential were under-appreciated and that the world was witnessing the "inevitable ascent" of a future world power. China, he declared, would move beyond its dominance in labour-intensive and mid-technology industries to push into knowledge-intensive areas that would "drive the future world economy" and "catapult China into the ranks of leading economic powers".[85] Shenkar argued that China's imperial past provided clues to its future course, causing it to pursue "an extremely high level of ambition". It would seek to restore its position as "a – if not *the* – leading civilization". He attributed China's exceptionality to "a special legacy, different institutions, sky-high ambitions, and a one-of-a-kind combination of resources, capabilities, and bargaining power". Its advance would not be linear, but it would be "unequivocally pointed forward". Moreover, it would become "a broker and arbiter of global diplomatic affairs not only in Asia but the world over". Rising military prominence would allow China to counterbalance the Western-dominated world "redrawing the political and security map and turning what it sees as a single player hegemony into a two- or three-player game".[86]

Despite his aggressive optimism, Shenkar was willing to consider other possibilities. He accepted that forecasting was "a tall task under any circumstances", and especially in the case of China. He was concerned that an economically belligerent China might arouse protectionist sentiment, although he thought this unlikely and that "a doomsday scenario" was more likely to arise from numerous internal problems. But these possible obstacles notwithstanding, he remained confident the twenty-first century would see China re-emerge as the Middle Kingdom and become a leading force in an interdependent world. It might "not only turn our economic theories and political assumptions on their head, but also test fundamental threads in our society".[87]

Shenkar's book was only the first of several that highlighted China's increasing global impact. Ted Fishman was equally impressed with China's economic dynamism. He noted that expert opinions about China's prospects had been "all over the map", producing "a combustible mix of euphoria, fear, admiration, and cynicism". But he felt confident that if any economy was going to supplant the United States, it would be China. In the not too distant future it might well have "the brand names, entertainment industry, and technology that set the world's standard". To Fishman, the dreams of China's hundreds of millions of new entrepreneurs had become "the most powerful force in the world".[88]

During the second half of the 2000s, several experts produced very optimistic forecasts for China's economic future. Robert Fogel calculated that in 2040 its economy would reach US$123 trillion, measured by purchasing power parity (PPP). This would be almost three times the GDP of the US and 40 percent of total global GDP.[89] Some argued China would become unstable unless it permitted competing political parties, reduced income inequalities, addressed environmental hazards, improved health programs, and countered corruption. However, Fogel thought the difficulty with these arguments was that they assumed its leaders were unaware of these problems, whereas in fact they were quite aware and were taking appropriate measures.[90]

Another US economist Albert Keidel was also bullish about China's economic prospects. He projected that China would overtake the US economy, in dollar terms, by 2035 and almost double it by 2050. In PPP terms, it would overtake the US in 2020 and almost double it in 2040. He assumed China's real growth rate would decline only gradually from 9.5 percent in 2010 to 5.5 percent in 2050.[91] Keidel expected that with an economy double the size of the US China would assume "global commercial and institutional leadership", including leadership of international organisations such as the United Nations, the World Bank, and the International Monetary Fund (IMF). The scope for unilateral action by the US would be increasingly constrained. He thought that later in the century China could also, if it wanted, become "a major – and possibly the leading – global military power". Keidel believed China was well able to cope with future challenges. It had evolved "a maturing, increasingly sophisticated decision-making system that is results-oriented". Its economy had become "a multi-tiered moneymaking machine". Many of its challenges were similar to those already faced and overcome by other fast-growing Asian economies.[92]

Optimistic perceptions of China's possible future influence were further boosted by the publication in 2009 of another best-selling book, this time by British analyst Martin Jacques with the attention-grabbing title *When China Rules the World.* He assessed that the hegemonic power of the United States was in decline – even though there was not yet widespread recognition of "the extent and irreversibility of that decline". In this new context he believed China's challenge to Western-style institutions and values would change the world profoundly. Jacques' underlying argument was that China's impact on the world would be as great as that of the US and "probably far greater". Given that China promised to be "so inordinately powerful and different", he found it "difficult to resist the idea that in time its rise will herald the birth of a new international order". In the long term it would sponsor "a new China-centric international system which will exist alongside the present system and probably slowly begin to usurp it". The US would "bitterly resist" this, with the consequence that "any transition will inevitably be tense and conflictual". He speculated that "as the balance of power begins to shift decisively in China's favour" the world might be divided into two spheres with East Asia and Africa under Chinese tutelage and Europe and the Middle East under the American umbrella.[93]

Somewhat surprisingly, in view of these clearly stated conclusions, in the body of his book Jacques gave considerable credence to the sceptics' views that there would be many potential obstacles to the sustainability of China's economic growth. Continued rural-to-urban labour migration might sustain its economy "for at least the next five to ten years, perhaps longer", but there were no guarantees. There were "powerful reasons" for believing China's current growth model was unsustainable. It was over-dependent on investment and income inequality was creating growing social tensions.[94] Jacques risked further undermining the credibility of his optimistic viewpoint by admitting that his approach rested on assumptions that "might prove to be wrong". He conceded that China's economic growth would "certainly decline" and thought it also likely there would be profound political changes, perhaps including "either the end of Communist rule or a major metamorphosis in its character". He continued to maintain, however, that it was "destined to become one of the two major global powers and ultimately *the* major global power". In his final chapter he explicitly stated that he assumed China's rise would not be derailed. He added that what would demolish his argument was if for some reason it "implodes"; but he thought this unlikely, even though it could not be excluded. Instead he expected China would continue to grow stronger and ultimately emerge over the next half-century or less as the world's leading power.[95]

Jacques was ambivalent about the future international posture of a Chinese superpower. On the one hand he believed it would begin to display "a millennia-old sense of superiority" and acquire "the interests and instincts of a superpower". But on the other hand (and rather incongruently with his book's title), he also believed that in an important sense China "does not aspire to run the world because it already believes itself to be the centre of the world". Hence it might be "rather less overtly aggressive than the West has been", but that did not mean it would be "less assertive or less determined to impose its will". Jacques believed that China's "global hegemony" was likely to grow over the next half century with traumatic consequences. The West would discover "that the world is no longer Western".[96]

In his concluding remarks Jacques observed that the prospect that China would fundamentally transform the world had been "consistently downplayed". It was "the elephant in the room that no one is quite willing to recognize". He believed that a country that saw itself "for both cultural and racial reasons, as the greatest civilization on earth" would "clearly in time require and expect a major reordering of global relationships". All this was well into the future because, for the next 20 years or so, while China modernised, it would "remain an essentially status quo power". However, in the longer term he was concerned about the consequences of China's "overweening sense of superiority and the hierarchical mentality this has engendered". But, again qualifying his earlier assessment, he argued that the decline of the West would not be followed simplistically by a Sino-centric world because China's rise was only "part of a much wider phenomenon, an increasingly multipolar world and the proliferation of diverse modernities". The new world "at least for the next century" would not be Chinese in the way that the previous one was Western, but China would "increasingly be in the ascendant and eventually dominant". Jacques closed with a brief comment that an unknown for China was the effect of its growth rate falling below 8 percent, and perhaps to 6 percent or lower, which could present "the biggest test Chinese society has faced since 1989". But he added that if it continued to grow at 6 to 8 percent, while Western economies entered negative or near zero growth, this would significantly accelerate China's ascendancy.[97]

Jacques' book provoked a vigorous riposte from Minxin Pei in an article titled *Why China Won't Rule the World.* Pei noted that predictions of Chinese dominance seemed to impress everyone but China itself. China's leaders were, he believed, more concerned with domestic stability than with projecting itself on the international stage – because they understood how difficult China's domestic challenges were and "how quickly fortune can turn".[98] The most trenchant review, however, came from Andrew Nathan who noted that, despite his "breathless title", Jacques actually seemed unsure of China's prospects. His "zigzag logic" left readers wondering whether China would "rise or founder". Nathan also objected to the "overheated and ofttimes downright scary language" employed by Jacques to predict "a coming reversal of dominance" between the West and China. Most of these predictions had "already happened" and were not cause for "histrionic alarm". He observed that Jacques' "on-path predictions" were the safest to make because "most of the time things do move in the same direction", but they did not do so "forever". Sooner or later, Nathan believed, "something unpredictable will occur" – for example, "democratization or system collapse".[99]

Despite such criticism, by mid-2011 Jacques' optimism seemed if anything to have grown. He saw China's economy overtaking the US by 2020 if not earlier. The *renminbi* would become the dominant world currency, the IMF and World Bank would be marginalised by China's influence, and Shanghai would replace New York as the global financial centre. China would achieve global hegemony in a world shaped by itself rather than by the US. Jacques believed this "Chinese world" needed to be understood and embraced because "willy-nilly … it will come to pass".[100] Again, in 2012, he predicted that within two decades China would steadily usurp the US as "the dominant global power".[101]

Collectively Shenkar, Keidel, and Jacques probably mark a high tide in expert enthusiasm for the idea of China's likely future dominance in world affairs. The optimism of many of their fellow-experts was frequently less fulsome and more qualified by awareness of the massive challenges China confronted. But, although more cautious, the views of many of them could still readily be described as optimistic. For example, from the mid-2000s the notion of China's possible fragmentation and disintegration became increasingly discredited. Barry Naughton and Dali Yang remarked that, since China had in fact been able to sustain national unity, it was more relevant to examine the centripetal forces that were holding it together than those that were allegedly pushing it apart.[102] Edward Friedman was also more confident in the ability of China's authoritarian regime to maintain internal stability. He found it possible to imagine bursting economic bubbles, tensions over Taiwan, leadership infighting, and "a more militaristic and fascistic direction"; but on balance thought that "at a great cost of gratuitous human suffering" China would hold together and continue to rise.[103] The assessment of another edited book by China scholars was that there were forces in China powerful enough to tear the country apart, but it concluded that the collapse of neither China nor the CCP was likely.[104]

One expert who saw even more reason to be positive was former US ambassador Charles Freeman. He thought it "not far-fetched" to consider that by 2025 China might have the world's largest economy, and rule by law might have brought about "a high level of domestic predictability and tranquillity". Habits of consultation and transparency, characteristic of democracy at its best, might have become integral to the CCP. China and the US might be leading the rehabilitation of the global environment and establishing a presence on the moon; and a new global concert of major powers might be bringing about "a harmonious and largely peaceful world order". Freeman recognised there were "many far darker scenarios than these", but warned that all that was required to realise them was "to behave as if they are inevitable and to interact with China as though they were". He argued there was nothing more inevitable about pessimistic outcomes than about brighter possibilities.[105]

The financial services firm UBS's Jonathan Anderson was another who projected a positive outlook for China, and backed this up with some well-structured arguments. He noted that "the myth of China's collapse" as typified by Gordon Chang was "wearing thin" and asked: "So which is it? Is China the world-beating juggernaut … or a fragile, stagnating economy on the verge of disintegration?" He thought the answer was neither. Most claims about China's looming economic supremacy were "egregious overstatement", but it was also "far from collapsing". Anderson noted that if you wanted to portray China as "a basket case", this could be supported by relevant data; but if, on the other hand, you wanted to trumpet its advantages, this could also be supported. He favoured the latter position and was adamant that the economy was "*not* heading for disaster". He thought China fitted exactly the Asian growth model and should have little problem sustaining average real growth of 7.5 to 8.5 percent over the following two decades. Noting that China's rise to date had been breathtaking, he nevertheless thought it fair to say: "you ain't seen nothing yet". By 2025 China's economy would be the world's largest.[106] In a debate with Minxin Pei, Anderson congratulated him on "an engaging review of long-term challenges", but denied he had made the case for a looming crisis. China's rise was "not some hazy prospect shimmering on the distant horizon, but a concrete reality". It did not need 10 percent growth – 8 percent would "do nicely" and even 6 or 7 percent would only push back the target by a few years. Economic failure would not occur without "a massive disturbance or outright crisis that derails growth for a long, long spell – and you need it fairly soon".[107]

Meanwhile, a consistently positive appraisal of China's likely future prospects had been upheld over a long period by Australian economist Ross Garnaut. He played a central role in meetings of professional economists from Australia, China and elsewhere, who gathered at the Australian National University for annual *China Update* conferences. Garnaut initially – in 1999 – took a cautious stance, noting that the modernisation of China's economy was such an immense undertaking that it would be "inevitably challenged" and would not proceed "in logarithmic straight lines". Nevertheless he believed it unlikely that the risks it faced would lead to economic collapse. It would be capable of returning to its growth path "whenever it is temporarily knocked from it". He considered that it would be "a natural and in no way a miraculous outcome" if growth continued until China's living standards rivalled those of the world's most advanced economies. He forecast that economic output would double each decade and that by 2020 China would be the world's second largest economy in dollar terms.[108] Few others at this time shared this degree of optimism, even though it proved an underestimation when China became the second largest economy in 2010. Garnaut later argued that China's economy would catch up with other countries sooner than most would expect, because the use of market exchange rates (MER) to compare national incomes undervalued the incomes of developing countries.[109] He was more concerned about its political and environmental situation.[110]

In 2005 Garnaut added that predictions of China's inevitable failure had been shown to be "already wrong". The pessimists and other "legions of commentators" who said its economy would not succeed were "scrambling out of the dustbin of history". Even the optimists had been "too timid in calling the scale and speed" of its growth. He acknowledged that China's economy was under "immense stress". The "wiser optimists" had always recognised there would be "bumps in the road, false turns … and detours". Nonetheless he thought there was reason for unprecedented confidence that it was on course to catch up with the world's advanced economies.[111]

Another Australian who, on balance, offered a positive prognosis for China's internal development was Hugh White. He expected its economy would "keep growing fast for another few decades" even if there were "bumps along the way". He noted, however, that because of political and social factors there was a real chance it might "stumble" – but equally it might not. He thought none of China's problems were "a certain show-stopper". White thought China's international influence would continue to increase, but did not believe it would "rule the world" because no matter how much it grew – even if it overtook the US economy – there would still be "too many other strong states to contend with".[112]

Optimistic projections by Chinese writers about their country's future have been equally as uncommon as the rare pessimistic projections noted earlier. But in one such example, in the early 2000s, two Chinese authors with both official and academic backgrounds pointed out that developments in contemporary China would have been unimaginable 20 years previously, and they suggested that changes 20 years hence would be "even harder to imagine". They believed there would be "no turning back" in China's modernisation, even though it confronted difficult challenges. They complained, not unreasonably, that numerous international strategic thinkers had failed repeatedly to appreciate that China's development potential far exceeded the constraints, and that "all sorts of lurking dangers" had already been put to rest. Predictions of China's collapse had therefore missed the mark "time and time again".[113]

Meanwhile, a new champion of Chinese optimism emerged. Hu Angang, a well-known economist and influential adviser to the Chinese government at Tsinghua University, has been conspicuously upbeat about China's prospects, providing possibly the most comprehensive forecast of China's future outlook by a Chinese author. In his 2011 book *China in 2020: A New Type of Superpower*, he stated that there was "no question" in his mind that China would indeed become a superpower. Moreover, it would be "a mature, responsible, and attractive superpower". He rejected "fallacies" such as that China was a bubble economy or that it might collapse. To become a superpower he reasoned that it needed to catch up with and surpass the United States in "economic power, human resources and human capital, science and technology achievements, and the ability to address climate change". He thought China was "trending in this direction" and that by 2020 it would attain superpower status, thereby ending the era of American hegemony and marking a shift "from unipolarity to a multipolar world dominated by the United States". However, he judged that this new era would be "defined less by competition than by cooperation" because the great powers would be increasingly interdependent and their shared interests would far outweigh areas of potential conflict.[114]

Hu considered that by 2020 China would have the world's largest economy (in PPP terms). It would also be the largest science and technology power, the global leader in innovation, "the largest environmentally friendly society", and "a new base for contemporary world civilization". However, he conceded that China's outlook was still "rife with uncertainty". It had already become "the superpower of natural resource consumption and greenhouse gas emissions" and would face mounting international pressure.[115] Expert views about such uncertainty are the subject of the next chapter.

Less uncertainty, and even greater optimism than Hu Angang's, was displayed in another and more recent book by fellow Tsinghua University scholar Yan Xuetong. In his book in Chinese titled *Inertia of History: China and the World in the Next Ten Years*, Yan argues that there is a powerful inertia behind current trends in China that will substantially enhance its influence and enable it, in the not too distant future, to play a more active international role. On current trends, Yan predicts that by 2023 China will probably have overtaken the GDP of the United States (at market exchange rates) and have also reached parity in comprehensive strength. It would become a second superpower in a new bipolar global structure, with strategic influence second only to the US. China's navy would also, by then, possess three to four aircraft carriers.[116]

Notes

1 Martin Jacques, *When China Rules the World: The Rise of the Middle Kingdom and the End of the Western World* (London: Allen Lane, 2009), 152.
2 John Fitzgerald, *Australia–China Relations 1976: Looking Forward*, R. G. Neale Lecture Series (The National Archives of Australia and the Department of Foreign Affairs and Trade, 2 August 2007), 10–12.
3 A. Doak Barnett, *Uncertain Passage: China's Transition in the Post-Mao Period* (Washington, DC: Brookings Institution, 1974), xi–xvi.
4 Ibid., 26–34, 112–118, 179–184 and 295–321.
5 Harry Harding, *China: The Uncertain Future* (New York: Foreign Policy Association, 1974), 3, 17 and 66–69.
6 Alan S. Whiting and Robert F. Dernberger, *China's Future: Foreign Policy and Economic Development in the Post-Mao Era* (New York: McGraw Hill, 1977).
7 Saburo Okita, "Japan, China and the US: Economic Relations and Prospects", *Foreign Affairs* 57, no. 5 (Summer 1979), 1103–1107.
8 Herman Kahn and the Hudson Institute, *World Economic Development: 1979 and Beyond* (Boulder, CO: Westview, 1979), 415–417.
9 Dwight H. Perkins, "The Chinese Economy in the 1980s", in *China Among the Nations of the Pacific*, ed. Harrison Brown (Boulder, CO: Westview, 1982). See also Dwight H. Perkins, *China: Asia's Next Economic Giant?* (Seattle, WA: University of Washington Press, 1986).
10 A. Doak Barnett, *China: Long-Term Development Issues and Options*, (Baltimore, MD: Johns Hopkins University Press, 1985).
11 A. Doak Barnett, "Ten Years After Mao", *Foreign Affairs* 65, no. 1 (Fall 1986), 63–65.
12 Gary Klintworth, *China's Modernisation: The Strategic Implications for the Asia-Pacific Region* (Canberra: Australian Government Publishing Service, 1989).
13 George Hicks, ed. *The Broken Mirror: China After Tiananmen* (Harlow, UK: Longman, 1990), xv–xx.
14 Ibid., 130 and 466–472.
15 Ibid., 149, 160, 194–195 and 455–465.
16 Roderick MacFarquhar, "The Anatomy of Collapse", *The New York Review of Books* (26 September 1991), 5–9.
17 W. J. F. Jenner, *The Tyranny of History: The Roots of China's Crisis* (London: Allen Lane, 1992), 1–2 and 227–248.
18 *China's Economic Dilemmas in the 1990s: The Problems of Reforms, Modernization, and Interdependence*, ed. Joint Economic Committee, Congress of the United States (New York: M. E. Sharpe, 1992), xiv.
19 Ibid., 8–13.
20 Ibid., 224–225.
21 This paragraph is based on analysis by Richard Baum, *Burying Mao: Chinese Politics in the Age of Deng Xiaoping* (Princeton, NJ: Princeton University Press, 1994), especially 20–21.
22 Chang, "China's Future: Regionalism, Federation, or Disintegration", 211 and 226.
23 Edward Friedman, "China's North–South Split and the Forces of Disintegration", *Current History* (September 1993), 270–274.

24 David Shambaugh, "Losing Control: The Erosion of State Authority in China", *Current History* (September 1993), 254–259.
25 Richard Hornik, "The Muddle Kingdom?: Bursting China's Bubble", *Foreign Affairs* 73, no. 3 (May/June 1994), 28–42.
26 Gerald Segal, "The Muddle Kingdom?: China's Changing Shape", *Foreign Affairs* 73, no. 3 (May/June 1994), 43–58.
27 David S. Goodman and Gerald Segal, eds. *China Deconstructs: Politics, Trade and Regionalism* (New York: Routledge, 1994), 350–351.
28 David S. Goodman and Gerald Segal, *China Without Deng* (New York: Imprint, 1995), 4–14 and 92.
29 Willy Wo-Lap Lam, *China after Deng Xiaoping: The Power Struggle in Beijing since Tiananmen* (New York: John Wiley & Sons, 1995), 383–430.
30 James A. R. Miles, *The Legacy of Tiananmen: China in Disarray* (Ann Arbor, MI: University of Michigan Press, 1996), 10–11 and 312–315.
31 Jack A. Goldstone, "The Coming Chinese Collapse", *Foreign Policy* (Summer 1995), 35–53.
32 Gerald Segal, "Does China Matter?", *Foreign Affairs* 78, no. 5 (September/October 1999), 24–36.
33 Callum Henderson, *China on the Brink: The Myths and Realities of the World's Largest Market* (New York: McGraw-Hill, 1999), xv–xxi and 243.
34 David S. Goodman and Gerald Segal, *China in the Nineties: Crisis Management and Beyond* (Oxford, UK: Clarendon, 1991), 17–18.
35 Ibid., 154–159.
36 *China's Economic Dilemmas in the 1990s: The Problems of Reforms, Modernization, and Interdependence*, 65–67.
37 Harry Harding, "On the Four Great Relationships: The Prospects for China", *Survival* 36, no. 2 (Summer 1994), 22–39.
38 Nicholas Kristof, "The Rise of China", *Foreign Affairs* 72, no. 51 (November/December 1993), 59–74.
39 Yasheng Huang, "Why China will not Collapse", *Foreign Policy* 99 (Summer 1995), 54–68.
40 Daniel Burstein and Arne de Keijzer, *Big Dragon: the Future of China* (New York: Touchstone, 1998), 19–21, 84, 97, 103 and 114.
41 Ibid., 341–348.
42 Chang's book title is reminiscent of Jack Goldstone's earlier article, cited above, and also of Brian Reading's *Japan: The Coming Collapse* (London: Weidenfeld & Nicholson, 1992).
43 Gordon G. Chang, *The Coming Collapse of China* (London: Arrow, 2002).
44 Ibid., xxv–xxvii.
45 Ibid., xv–xxix, 16 and 44.
46 Ibid., 251–279.
47 Robert Sutter, "The Chinese Regime will Endure", *China Brief* 1, no. 8 (25 October 2001).
48 June Teufel Dreyer, "What if Chang is Right?", *China Brief* 1, no. 8 (25 October 2001).
49 Gordon G. Chang, "China's Critical Moment", *China Brief* 1, no. 8 (25 October 2001). This period coincided with the timeframe predicted earlier by Goldstone.
50 Gordon G. Chang, "The Collapse of China, Act 1", *China Brief* 2, no. 3 (31 January 2002).
51 Gordon G. Chang, "Is this China's Century?", *China Brief* 3, no. 16 (16 December 2003).
52 Gordon G. Chang, "Collapse Perhaps: The Stability of the Modern Chinese State", (27 March 2004). www.gordonchang.com.
53 Gordon G. Chang, "Halfway to China's Collapse", *Far Eastern Economic Review* (June 2006).

54 Gordon G. Chang, "The Coming Collapse of China: 2012 Edition", *Foreign Policy* (29 December 2011).
55 He Qinglian, "A Volcanic Stability", *Journal of Democracy* 14, no. 1 (January 2003), 66–72.
56 Moises Naim, "Only a Miracle can Save China", *Financial Times* (15 September 2003). See also George Gilboy and Eric Heginbotham, "China's Coming Transformation", *Foreign Affairs* 80, no. 4 (July/August 2001), 26–39. Gilboy and Heginbotham saw China as "a non-adaptive brittle state" threatened by systemic collapse.
57 Minxin Pei, "Will China become another Indonesia?", *Foreign Policy* (Fall 1999).
58 Minxin Pei, "China's Governance Crisis", *Foreign Affairs* 81, no. 5 (September/October 2002), 97–101.
59 Ibid., 101–109.
60 Minxin Pei, "Dangerous Denials", *Foreign Policy* (January/February 2005).
61 Minxin Pei, *China's Trapped Transition: the Limits of Developmental Autocracy* (Cambridge, MA: Harvard University Press, 2006), 4–11 and 203–215. Andrew Nathan, in a perceptive review of Pei's book, commented that he had given "only weak evidence" to support his contention that China needed to adopt democracy and full marketisation. In Nathan's view the question of whether China could prosper under an authoritarian regime was "exactly what is at stake in the Chinese experiment". See Andrew J. Nathan, "Present at the Stagnation: Is China's Development Stalled?", *Foreign Affairs* 85, no. 4 (July/August 2006), 177–182.
62 Minxin Pei, "The Dark Side of China's Rise", *Foreign Policy* (March/April 2006), 32–40.
63 Minxin Pei, *Corruption Threatens China's Future* (Carnegie Endowment for International Peace, October 2007).
64 Minxin Pei, "China's Not a Superpower", *The Diplomat* (29 December 2009).
65 Minxin Pei and Jonathan Anderson, "The Color of China", *The National Interest* (March/April 2009), 13–19.
66 Minxin Pei, "Everything you Think you Know about China is Wrong", *Foreign Policy* (29 August 2012).
67 Will Hutton, *The Writing on the Wall: China and the West in the 21st Century* (London: Little, Brown & Co, 2006), x–xii and 1–20.
68 Ibid., 30–34, 118–120 and 331–357. In an interesting debate with British economist and politician Meghnad Desai, Hutton insisted China could only compete with the West if it became "more like us". See Will Hutton and Meghnad Desai, "Does the Future Really Belong to China?", *Prospect Magazine* (January 2007).
69 John Lee, *Will China Fail?* (St Leonards, NSW: Centre for Independent Studies, 2007), 4, 13, 96–99 and 112–118.
70 Ibid., 3–7, 17–18 and 146.
71 John Lee, "Beyond the Beijing Olympics: Reassessing China's Future", in *China and the Olympics* (Australian Institute of International Affairs, June 2008), 21–31.
72 Lee, *Will China Fail?*, 3–4, 14–29, 95–99, 112–146 and 151–155. Stephan Halper shared Lee's concern, although he was less inclined to question China's future prospects. He believed the CCP had made a "civic bargain" with the Chinese people, in which political repression was tolerated in exchange for stability and economic advancement. See Stefan Halper, *The Beijing Consensus: How China's Authoritarian Model will Dominate the Twenty-first Century* (New York: Basic Books, 2010), ix–xii, 2–3 and 207–252.
73 Despite Monk's scepticism, by 2005 China's economy had already expanded 20-fold since 1981 and seems likely to grow 50-fold before 2030 even with much reduced annual growth.
74 Paul Monk, *Thunder from the Silent Zone: Rethinking China* (Melbourne, VIC: Scribe, 2005), 48–62.
75 Ibid., 50–59.

76 Joshua Kurlantzick, "Asia Minor: Is China's economic boom a myth?", *The New Republic* (16 December 2002), 20–25.
77 Joe Studwell, *The China Dream: The Quest for the Last Great Untapped Market on Earth* (New York: Grove Press, 2003), xii, 286 and 292.
78 Peter Navarro, *The Coming China Wars: Where they will be Fought and How they can be Won* (London: FT Press, 2006), v, xii–xix and 199–217.
79 Wang Shaoguang, Hu Angang, and Ding Yuanzhu, "Behind China's Wealth Gap", *South China Morning Post* (31 October 2002).
80 Zhang Ruizhuang, "Would there be Two Giants Living in the same Mountain? The Geostrategic Implications of China's Rise for US–China Relations", in Eva Paus, Penelope B. Prime, and Jon Western, eds. *Global Giant: Is China Changing the Rules of the Game?* (New York: Palgrave Macmillan, 2009), 219–226.
81 Noy Siakhachanh, *China 2020: Challenges Ahead* (Deutsche Bank Research, 2002).
82 Chetan Ahya and Andy Xie, *India and China: New Tigers of Asia* (Morgan Stanley, 26 July 2004). Commendably, Morgan Stanley flagged that its investment advice created a potential conflict of interest that could affect objectivity.
83 Dominic Wilson and Roopa Purushothaman, *Dreaming with BRICs: the Path to 2050* (Goldman Sachs, 1 October 2003).
84 Dominic Wilson, Roopa Purushothaman, and Themistoklis Fiotakis, *The BRICs and Global Markets: Crude, Cars and Capital* (Goldman Sachs, 14 October 2004). Further updating was required in later years as China's economy returned to double-digit growth and outperformed even Goldman Sachs's more optimistic forecast.
85 Oded Shenkar, *The Chinese Century: the Rising Chinese Economy and its Impact on the Global Economy, the Balance of Power, and Your Job* (Upper Saddle River, NJ: Wharton School, 2006), 1–5.
86 Ibid., 25–43 and 161–165.
87 Ibid., 169–177 and 187.
88 Ted C. Fishman, *China, Inc: How the Rise of the Next Superpower Challenges America and the World* (New York: Scribner, 2006), 12, 16–17, 41, 252, 293–296 and 303.
89 Measurement of China's GDP in terms of purchasing power parity (PPP) increases the monetary value of China's GDP by taking into account the fact that a standard "basket" of goods and services can be produced more cheaply in an economy with lower costs and prices than in the developed world. However, GDP is more commonly measured in terms of current market exchange rates (MER) where GDP values are expressed in US dollars. The numerical value of a developing country's GDP is much higher in terms of PPP than at MER. Fogel's calculations were based on the doubtful assumption that China's economic growth would average 8.4 percent from 2000 to 2040.
90 Robert W. Fogel, "Capitalism and Democracy in 2040: Forecasts and Speculations", *Daedalus* (Summer 2007), 8–13.
91 Albert Keidel, *China's Economic Rise – Fact and Fiction* (Carnegie Endowment for International Peace, July 2008), 1–6.
92 Ibid., 6–13.
93 Jacques, *When China Rules the World*, 6–16 and 362.
94 Ibid., 163–174.
95 Ibid., 363–364.
96 Ibid., 392–394 and 409–413.
97 Ibid., 429–434.
98 Minxin Pei, "Why China won't Rule the World," *Newsweek* (8 December 2009). Publication of Jacques' book also led to a vigorous discussion with Will Hutton. See: Martin Jacques and Will Hutton, "Is Western supremacy but a blip as China rises to the global summit?", *Guardian* (23 June 2009).
99 Andrew J. Nathan, "The Truth about China", *The National Interest* 105 (January/ February 2010).

100 Martin Jacques, "China to Trounce US in Next Decade", *The National Interest* (7 July 2011).
101 Martin Jacques, "Why do we continue to ignore China's rise: Arrogance", *Observer* (25 March 2012). Jacques stated in an interview with *Australia China Quarterly* on 11 October 2012 that the title of his book was "ironic" and "not meant to be taken literally".
102 Barry Naughton and Dali L. Yang, eds. *Holding China Together: Diversity and National Integration in the Post-Deng Era* (Cambridge, UK: Cambridge University Press, 2004), 2–24. See also Zheng Yongnian, *De facto Federalism in China: Reforms and Dynamics of Central–Local Relations* (Singapore: World Scientific, 2007). Zheng maintained that a de facto federal system had evolved in China, with mutual accommodation between the centre and provinces. For Yang's earlier views see Dali L. Yang and Houkai Wei, "Rising Sectionalism in China?", *Journal of International Affairs* 49, no. 2 (Winter 1996), 456–457 and 475–476.
103 Tun-Jen Cheng, Jacques Delisle, and Deborah A. Brown, eds. *China Under Hu Jintao: Opportunities, Dangers, and Dilemmas* (Singapore: World Scientific, 2006), 98–99 and 129–133.
104 Czeslaw Tubilewicz, ed. *Critical Issues in Contemporary China* (New York: Routledge, 2006), 2–17.
105 Freeman, "China in the Times to Come".
106 Jonathan Anderson, *How to Think about China* (UBS Investment Research, January 2006), 3–4, 36–37, 77–78, 116, 147 and 194.
107 Pei and Anderson, "The Color of China".
108 Ross Garnaut, "Twenty years of economic reform and structural change in the Chinese economy", in Ross Garnaut and Ligang Song, eds. *China: Twenty Years of Economic Reform* (Canberra: Asia Pacific Press, 1999), 9 and 24–25.
109 Ross Garnaut, "Catching up with America", in Ross Garnaut and Ligang Song, eds. *China 2002: WTO Entry and World Recession* (Canberra: Asia Pacific Press, 2002), 5 and 11. Garnaut noted that it had become conventional to query China's growth statistics, but he believed them to be "not wildly misleading".
110 Ross Garnaut and Ligang Song, eds. *China: Is Rapid Growth Sustainable?* (Canberra: Asia Pacific Press, 2004), 11–21; and Ross Garnaut and Ligang Song, eds. *China: Linking Markets for Growth* (Canberra: Australian National University E Press, 2007), 1–3.
111 Ross Garnaut, "The China boom and its discontents", in Ross Garnaut and Ligang Song, eds. *The China Boom and Its Discontents* (Canberra: Australian National University E Press, 2005), xiii–xx.
112 Hugh White, "Power Shift: Australia's Future Between Washington and Beijing".
113 Wang Zhan and Huang Renwei, "What will the World Gain from China in Twenty Years", *China Business Review (*March–April 2003), 36–40.
114 Hu Angang, *China in 2020: A New Type of Superpower* (Washington, DC: Brookings Institution, 2011), ix, 2, 12–17. For a different view of China's superpower aspirations see Chu Shulong, "Can China be called an old-style superpower?", *Global Times* (22 July 2009).
115 Ibid., 22, 33–36, 95–110 and 157–158.
116 Yan Xuetong, *Inertia of History: China and the World in the next Ten Years* (*Lishi de guanxing: Weilai shinian de Zhongguo yu shijie)* (Beijing: China CITIC Press, July 2013).

Bibliography

Anderson, Jonathan. *How to Think About China*. UBS Investment Research, *Asian Economic Perspectives,* January 2006.

Barnett, A. Doak. *Uncertain Passage: China's Transition in the Post-Mao Period.* Washington, DC: Brookings Institution, 1974.

Baum, Richard. *Burying Mao: Chinese Politics in the Age of Deng Xiaoping*. Princeton, NJ: Princeton University Press, 1994.

Burstein, Daniel and Arne de Keijzer. *Big Dragon: The Future of China.* New York: Touchstone, 1998.

Chang, Gordon G. *The Coming Collapse of China*. London: Arrow, 2002.

Fishman, Ted C. *China, Inc: How the Rise of the Next Superpower Challenges America and the World.* New York: Scribner, 2006.

Goldstone, Jack A. "The Coming Chinese Collapse". *Foreign Policy* (Summer 1995).

Goodman, David S. and Gerald Segal. *China without Deng.* New York: Imprint, 1995.

Harding, Harry. *China: The Uncertain Future*. New York: Foreign Policy Association, 1974.

Henderson, Callum. *China on the Brink: The Myths and Realities of the World's Largest Market*. New York: McGraw-Hill, 1999.

Hicks, George, ed. *The Broken Mirror: China after Tiananmen*. Harlow, UK: Longman, 1990.

Hu Angang. *China in 2020: A New Type of Superpower*. Washington, DC: Brookings Institution, 2011.

Hutton, Will. *The Writing on the Wall: China and the West in the 21st Century*. London: Little, Brown & Co, 2006.

Jacques, Martin. *When China Rules the World: The Rise of the Middle Kingdom and the End of the Western World.* London: Allen Lane, 2009.

Keidel, Albert. *China's Economic Rise – Fact and Fiction.* Carnegie Endowment for International Peace, July 2008.

Lee, John. *Will China Fail?* St Leonards, NSW: Centre for Independent Studies, 2007.

Monk, Paul. *Thunder from the Silent Zone: Rethinking China.* Melbourne, VIC: Scribe, 2005.

Paus, Eva, Penelope B. Prime and Jon Western, eds. *Global Giant: Is China Changing the Rules of the Game?* New York: Palgrave Macmillan, 2009.

Pei, Minxin. *China's Trapped Transition: The Limits of Developmental Autocracy*. Cambridge, MA: Harvard University Press, 2006.

Shenkar, Oded. *The Chinese Century: The Rising Chinese Economy and its Impact on the Global Economy, the Balance of Power, and Your Job.* Upper Saddle River, NJ: Wharton School, 2006.

Studwell, Joe. *The China Dream: The Quest for the Last Great Untapped Market on Earth.* New York: Grove Press, 2003.

Tubilewicz, Czeslaw, ed. *Critical Issues in Contemporary China.* New York: Routledge, 2006.

White, Hugh. "Power Shift: Australia's Future between Washington and Beijing", *Quarterly Essay* 39 (2010).

Whiting, Alan S. and Robert F. Dernberger. *China's Future: Foreign Policy and Economic Development in the Post-Mao Era.* New York: McGraw Hill, 1977.

Yan Xuetong, *Inertia of History: China and the World in the Next Ten Years (Lishi de guanxing: Weilai shinian de Zhongguo yu shijie)*. Beijing: China CITIC Press, July 2013.

3 Acknowledging uncertainty

A great divide has characterised much of the expert debate about China for more than two decades. Neither pessimists nor optimists seemed willing to draw back from their entrenched positions. They often displayed, in their inflexible defensiveness, many of the characteristics of Tetlock's hedgehogs. It is important to point out, however, that although a clear polarisation of expert opinion did emerge, there were nonetheless a number of pessimists and optimists (perhaps especially the latter) who were less fervent or unequivocal than others. The previous chapter has revealed that many of them were willing to admit the contingent nature of their judgements and even the possible merits of contrary views.

Moreover, many other experts cannot be clearly labelled as either pessimists or optimists, no matter how open-minded. They cannot be categorised in either of these groups because they accepted much more readily the extent of fundamental uncertainty about many aspects of China's future. This often led them to pursue other approaches that permitted them to acknowledge and confront this uncertainty. Like Tetlock's foxes they were frequently more empirical, less predisposed to one argument or another, and ready to consider other viewpoints. They were more concerned about expert subjectivity and bias, and about excessive reliance on intuition, ideological agendas, theoretical assumptions, or on a single big idea. They displayed greater humility in the face of uncertainty and were more willing to reassess and change their views over time. They included many experts who came to believe that China's prospects are essentially uncertain, unpredictable, or dependent on evolving circumstances. The views and approaches of such experts are considered in this chapter.

Expert responses to uncertainty

One such expert was Avery Goldstein. He observed that forecasting was "fraught with uncertainty", to the extent that predictions about rapidly changing systems such as China's might be "an exercise in futility". It was difficult when analysing "smaller pieces of the big picture" to identify the most important forces and their likely consequences. Deductions that were almost sure to be wrong included those that relied on "simplistic historical analogies" or on "excessive confidence in the indeterminate implications of social science theories". He wished his own assessments to be considered not as specific predictions but as "a modest attempt to highlight the key issues and forces".[1]

The difficulty of assessing China's future has perhaps been most eloquently captured by veteran China watcher Orville Schell. He commented that "the more one knows about China ... the more difficult it is to assess its successes, much less prognosticate with any certainty about its future trajectory". Noting the division between those who were either impressed or sceptical about China's prospects, he suggested that in truth "both contradictory versions of China are real". This was so because China was "in the process of one of the most tectonic and monumental transitions of any nation in the last hundred years". He admitted that even China specialists with many years of experience were confused by the fact that it "embraces within it contradictions that can lead even an astute observer to equal and quite opposite analyses of its health and its future". He speculated that China's stunning development might have come with "a hidden price which will only slowly come due". In sum, he thought Mao's view that the best way to analyse a situation was through "the lens of contradiction" was still relevant for contemporary China, and "perhaps the only way to make sense of it".[2]

In a fascinating interview in 2001, Schell explained that what made China difficult to prognosticate about was that opposing scenarios were "almost equally plausible". He could argue the collapse scenario and be "utterly convincing", and also "quite compellingly" argue the peaceful evolution scenario. He believed all one could say about China was that it was "incredibly unresolved", more so than any other country of consequence. He stood in awe of China's ability to unleash "very dynamic, tectonic market forces" and still be able to maintain political control. But the million-dollar question was whether this would last. He thought it even "somewhat counterintuitive" that China could resolve itself in a tidy fashion. In 1989 he would not have thought China could come so far, but he still believed it would be very difficult for the CCP to overcome China's "very, very severe and ever tightening contradictions". He wondered how long China could go "without an economic downturn, an environmental catastrophe, a famine, a flood, a war, something?" A global downturn would also "really mess things up" because growth was the CCP's "only source of real legitimacy". Many Chinese were apparently happy with authoritarianism, but there also seemed to be "an ineluctable urge for people to have transparency, openness, a voice and participation".[3]

At times Schell appeared to lapse into pessimism. He accepted that even the most sceptical critics were "amazed at China's developmental prowess", but considered that no other country of global consequence was "more in a state of irreconcilable contradiction". It was in the midst of "a fevered but precarious transition". He feared that a destructive collision between the CCP and various centrifugal forces seemed "almost inevitable". China's economic miracle was built on "a pyramid of dreams". If its economic fundamentals weakened, a collapse of confidence might be "frighteningly rapid and devastatingly destructive".[4]

Over the course of the past decade many other experienced China watchers have echoed Schell's sentiments. They acknowledged equally both the strengths and the weaknesses of a rising China. Kenneth Lieberthal thought that, if it effectively managed a number of major challenges, China could in the three decades to 2033 become the world's second most powerful country. But he cautioned that unpredictable obstacles were bound to arise, and that the "realistic array of possibilities" for China's future was sobering.[5] Willy Lam perceived China as "a bafflingly complex country" and a "bundle of contradictions". There were in fact two Chinas: the China of the optimists, which was "an economic juggernaut" set to overtake the United States by 2030; and a China whose "overstretched social fabric" was "close to breaking point", where even a minor mishap could "ignite a full-scale explosion". Lam cited a Chinese Academy of Social Sciences (CASS) sociologist who thought China was at a crossroads where it could either evolve smoothly or "spiral into stagnation and chaos".[6]

Andrew Nathan also recognised both sides of the argument. He criticised Minxin Pei's "purely negative account" for not presenting a balanced picture of China's condition. It was taking steps to address its governance deficits and tolerated a rich internal debate about "the regime's pathologies". Pei also failed to take into account what Nathan saw as "the central fact of Chinese political life today: the power and unity of the central party elite". But Nathan added that "even the most astute analysis of current trends may not accurately predict the future". Major societal changes were "often disjunctive", hence it was prudent to consider potential "wild cards" – such as a drastic weakening of the US economy, or a war between China and the US over Taiwan – that could derail China's progress.[7] David Lampton, for his part, perceived "a marked paradigm shift" in expert thinking about China that threatened "to substitute one flawed framework (a 'weak China') for another (a 'China on steroids')". The reality in his view was that China was "weak and strong simultaneously". It was becoming a global economic force, but confronted huge developmental challenges that justified "reserved predictions" about its prospects.[8]

Journalist James Kynge was equally if not more cautious, observing that the rise and decline of great powers could rarely be traced in a simple, linear fashion but was "full of twists and turns, false dawns and deceptive signals". He was impressed by the emergence of China as "a palpable global force". To the surprise of the rest of the world, an authoritarian one-party state had succeeded in fostering a vibrant capitalist economy. However, predictions that China would become the world's most powerful economy might be flawed because the process by which China was turning back the clock was "full of human suffering, alienation and longing". China was also: "like an elephant riding a bike. If it slows down, it could fall off". But he hoped that China's pragmatism would supply a counter-argument to gloom and doom scenarios.[9] Joseph Fewsmith thought the question was whether the Chinese model could be modified incrementally or would be bent to the breaking point, and if the latter was the outcome then change could be rapid and unpredictable.[10]

George Walden likened China's journey to a trip on the Shanghai Maglev train where the passengers "gaze at the scenery zooming past part in wonderment, part in apprehension". He proposed that observers outside China should emulate them, marvelling at the ride but also "alert and apprehensive".[11] Uncertainty also marked British analyst Kerry Brown's views about China's prospects, although he sometimes tended toward pessimism. He was confident China would continue to advance, but also impressed by the scope for serious setbacks. He wondered whether, looking back on China in ten or 20 years time, the mid-2000s might be seen as "the period of over optimism", comparable to earlier optimism about Japan.[12] If it could manage its internal transition China could become one of the dominant global powers, but this was "a big if".[13]

Diversity of opinion among China experts was clearly evident in a stocktaking exercise undertaken in 2008 by *China Security* journal, which canvassed opinions about China's future from a large and disparate group of "leading thinkers" and indicated substantial ambiguity and confusion about China's outlook. The Editor likened this experiment to the famous story of blind men investigating an elephant, with each one having a quite different perception depending on which part of the elephant they encountered. Comprehending such a rapidly changing country as China seemed "a similarly confounding task". If any consensus was discernible from this survey of opinions, it indicated "deep uncertainty" about China's prospects.[14]

The challenge of uncertainty was eloquently addressed in a 2011 book edited by David Shambaugh. As he put it: "The only thing that is completely predictable about China's future is its unpredictability." The forecasts of some sinologists had proved "insightful and reasonably prescient" but many had foundered, often due to "mistaken attempts to impose logics, ideologies and analyses developed outside, or templates developed for the study of other systems and societies". China was "not completely *sui generis*" but it had "its own intrinsic dynamics". Contributors to the book, Shambaugh stated, were all appropriately cautious in their forecasts. None predicted "cataclysmic change" in the short- to medium-term, but they were not complacent. They anticipated a "dynamic stability". Should this "*evolutionary* rather than revolutionary perspective" be proved wrong, they risked subsequent ridicule by those of their colleagues who had been predicting "wrenching and systematic change".[15]

The expert opinions canvassed above confirm that the views of pessimists and optimists did not constitute the full spectrum of opinion about China's future. The middle ground was well populated by those who accepted the existence of considerable uncertainty. The remainder of this chapter considers several approaches used by experts in their attempts to manage the problem of uncertainty – including highlighting possible disruptors of China's progress, using expert groups to promote a greater degree of consensus, and addressing uncertainty by considering alternative outcomes through scenario construction.

What could go wrong?

Some experts chose a new way to address the issue of uncertainty. Putting the positives to one side, they focussed instead on the potential disruptors. They concentrated on the question: "What could go wrong?" This was an innovative approach which had the merit of not prejudging China's prospects. In practice, however, although most of these experts were more open-minded about China's prospects than the pessimists, they also tended to emphasise the likelihood of negative outcomes. Most were doubtful that China's rise could proceed uninterrupted given the abundance of challenges it would be likely to face.

In an early and ambitious example of this approach, Charles Wolf led a team at the RAND Corporation in the early-2000s that provided what one reviewer described as "a usefully sober alternative to the familiar rosy scenario". The report sought to identify and assess the "potential adversities or fault lines" that might cause China's economy to "go seriously awry" during the following decade. Potential fault lines were examined in eight likely problem areas: unemployment, poverty and social unrest; corruption; epidemic diseases; water shortage and pollution; energy; financial fragility and state-owned enterprises; foreign investment; and Taiwan and other potential conflicts. The report attempted to assess how severely China's economy would be affected if these fault lines impacted "separately or in clusters". It claimed to offer a countervailing perspective to an allegedly "generally prevailing consensus, with a few notable exceptions," that China could maintain high growth indefinitely.[16]

Wolf and his team attempted to quantify the possible negative impact on China's GDP of each potential fault line. They estimated that this impact could range from a reduction in GDP of 0.3 to 0.8 percent in the case of unemployment, poverty, and social unrest, to a reduction of 1.8 to 2.2 percent in the case of epidemic diseases. These percentages were calculated on the basis that each fault line occurred in isolation. The probability that *none* of these individual setbacks would occur was deemed to be low and that *all* would occur together lower still, but it was considered likely that if one or two adversities occurred the cross impacts would raise the probability that some of the other adversities would also occur. The authors of this intriguing method conceded that their results begged the question of how long each of the adverse effects would endure without inducing remedial measures.[17] The project displayed commendable creativity, although the credibility of its attempt to quantify the likely impacts on GDP for each of the fault lines was certainly open to question.

Another paper that concentrated on negative trends was produced by June Dreyer. She discussed the potential impact on China's growth of possibly underestimated population growth, growing disparities between rural and urban incomes, inadequate social security, high levels of unemployment, an unstable financial system, widespread corruption, adverse effects of globalisation, energy constraints, environmental destruction and water shortage, as well as international issues such as territorial disputes. She considered that these factors, which interacted with subtle and unpredictable effects, could jeopardise China's development. Although she did not see any reason to predict imminent collapse, she thought negative trends were likely to prevent China from "becoming the new hegemon, or even a superpower, any time soon".[18] Similarly in an article titled "What could go Wrong with China?", financial journalist Jonathan Laing produced another long list of things that could indeed go wrong, concluding that the consequent risk of political instability and social unrest made China's future unpredictable.[19]

Former US Deputy Assistant Secretary of State Susan Shirk also emphasised the fragility and vulnerability underlying China's apparent success. Like others, she highlighted a long list of domestic risk factors. In particular, China could be "ambushed by a political or economic crisis". From the outside it looked like a powerhouse, but to its leaders it seemed to be "overwhelmed by internal problems". She saw China as a brittle, authoritarian regime with "a deep sense of domestic insecurity". Although it was an emerging superpower, it was therefore a fragile one, whose progress could easily be upset.[20]

Another analyst who focused on what could go wrong was Wing Thye Woo. He sought to identify the "high-probability challenges" to China's continued high growth. Comparing China's economy to a speeding car, he identified three types of failure that could lead to a crash, namely a *hardware* failure, a *software* failure, and a *power supply* failure. The most likely hardware failure, he predicted, would be economic breakdown – comparable to a broken chassis. The most likely software failure would be a flaw in governance that created widespread social disorder – comparable to a fight among the occupants of the car. He believed the most likely power supply failure would be "a natural limit or an externally imposed limit" – comparable to running out of fuel or having the engine switched off.[21] In a later analysis, Woo agreed with Minxin Pei's portrayal of a trapped transition, but acknowledged that enlightened self-interest might triumph.[22]

Nicholas Eberstadt also questioned whether China's ascendency was assured. He thought it worthwhile to consider "other plausible possibilities", including economic, political, or military risk factors that might "complicate, hinder, or entirely forestall" China's rise. He observed that the difficulty with assessing alternative futures was "in having any presentiment today that such seemingly unlikely visions might actually unfold tomorrow". But, in his opinion, there was one problem area that had "a high degree of regularity and even predictability", namely China's aging demographic structure, which he claimed, was a "largely neglected" and "clearly unfavourable" factor.[23]

The consequences of things going wrong were also considered by David Lampton, who saw at least three paths that could lead China to fail, and if one of these began to unfold it would probably "cascade into the others". First, various issues could lead to economic disaster; second, there could be a political breakdown; and, third, China could reach its ecological limits. Nevertheless, he believed China would probably continue to enjoy vigorous growth for "the next couple of decades".[24] Other experts were not so confident about China's prospects in the face of its multitude of problems. There appears to be a growing school of thought that at the end of the day these problems are so diverse and so serious that it is unlikely that China will be able to avoid being seriously set back by one or more of them.

Expert groups

Up to this point, this chapter has concentrated on the forecasting techniques used by prominent individual experts to address the issue of uncertainty. However, contemplation of China's future has naturally not been the exclusive preserve of individuals. Within and across international organisations, business, government, academia, research institutes, and think tanks, there has been extensive collaboration *between* experts in joint efforts to capitalise on combined skills and knowledge. It is notable, however, that even in these cases there is relatively little evidence of serious efforts to combine expert judgements and to reach consensus. There have certainly been many, many instances where experts have come together to share their expertise – through conferences and edited books, debates and discussions, and collaborative research and analysis – but relatively few cases at least in the public sphere where this has resulted in documented attempts to reconcile the conflicting views of individual experts.[25] The exceptions to this rule were important, however. As the following summary indicates, these exceptions usually resulted in assessments that emphasised the high degree of uncertainty in China's outlook.

It is perhaps surprising that one of the most concerted attempts to shape a consensus of opinion among a group of China experts was also one of the earliest, occurring in the period immediately before Mao's death. This took the form of a Delphi study coordinated by Czech–Canadian analyst Vaclav Smil. Fifty experts from mainly Western countries were invited to participate, and 20 accepted. The exercise involved three rounds of questioning between November 1974 and March 1975. The experts were asked to give their views on a consensus list of the most "critical problems" affecting China's future between 1975 and 2000, and on the likelihood and probable timing of 20 different major events. Of these events the most likely was assessed to be the emergence of a new "revisionism" in which revolutionary radicalism would be replaced by a more pragmatic style of governance. The chance of a leadership succession crisis was listed second. Interestingly, in view of forecasts in later years, the prospect of regional fragmentation, and of a collapse of the CCP, were among the lowest ranked. In Smil's opinion, Delphis were "just an exploratory tool" with obvious limitations, but they were useful where there was a high degree of uncertainty and disagreement. He thought criticisms of this technique often misrepresented it.[26] Delphis seem now to be out of fashion, having perhaps become excessively and unnecessarily complex and exacting in their application. But it seems regrettable, especially given the relative success of Smil's Delphi, that no other such exercises focussing on China have ever been attempted.

One of the biggest supporters of efforts to bring experts together to forecast China's future has been the United States government. Some of this research, and especially that of the RAND Corporation, has made its way into the public realm. In 1995, for example, a RAND project attempted to identify indicators of adverse change in the Asia–Pacific region over the following 10–15 years. The resulting report, coordinated by Michael Swaine, was supported by "the labours of many individuals" including US scholars and officials. It was cautious in many respects but nonetheless came to some quite bold and specific conclusions, often contradicting the pessimistic views of previous authors. For example, it correctly forecast that domestic changes in China would "almost certainly not produce a significantly more democratic and pro-Western Chinese regime". Neither would they "lead to the emergence of independent regional power centers or the complete breakdown of political rule in China". The report predicted accurately that China's authoritarian government and its current foreign policy stance would probably "continue for many years … marked by overall caution and pragmatism, a recognition of the need for a placid regional environment to permit a continued emphasis on economic reform, and a balancing of both cooperation and competition with the West". Offsetting these rather positive conclusions, the report highlighted the influence of the Chinese military and of "ultraconservative nationalistic sentiments".[27]

Two years later, in 1997, a joint study by RAND and the Pacific Council also included participants from diverse professional backgrounds. They met eight times – four hours each session – and to their surprise reached a consensus that they did not see China "as an inevitable and implacable enemy, nor as a country likely to collapse", but they also stressed the multiple uncertainties China faced. They rejected pessimistic views expressed "with increasing frequency" in the US that China was a totalitarian and oppressive country, that it clearly intended to become "the dominant power in Asia and possibly the world", that it would be unable to reform itself, that its disintegration had already begun, and that the regime was "heading for the dustbin of history". The study also rejected an "optimistic view" that China's inevitable economic growth would "propel it towards political liberalization and ultimately democratization". It stressed the uncertainties, vulnerabilities, risks, openness, and complexities of China's future. It also spent some time demolishing what participants called the "conventional wisdom" – which ironically rather closely resembled the actual outcome – that the Chinese economy was "on a trajectory of rapid growth likely to last for many more years, perhaps decades", and that it was "a looming power and rival to the United States". Straight-line projections that China would continue to grow at nearly 10 percent a year were considered unlikely.[28]

Another significant attempt by US experts to reach a consensus about China's future prospects occurred in the context of a conference in 1998. Participants adopted the unusual approach of concentrating on the various factors that either contributed to, or detracted from, stability in China. Some expressed quite negative views. David Shambaugh thought that the systemic conditions for regime decay or collapse were present. Bruce Dickson thought a revolutionary situation could develop "quickly and unexpectedly" and that fundamental political change could then be "sudden and tumultuous". Dorothy Solinger, on the other hand, believed the coercive power of the state could be brought to bear to contain social unrest. H. Lyman Miller thought the CCP was capable of taking the initiative and being "the main author and the conscious agent" of change. The majority judged that factors favouring stability were dominant for the present, but that no reasonable analyst should assume this would continue indefinitely. Any predictions should be relative and contingent because they had "all too often proven wrong".[29]

Yet another assembly of US experts, who described themselves as the Aspen Strategy Group, met in the mid-2000s. They included senior figures from government and business as well as participants from China. A disparate range of views emerged, many of which were quite negative. Wang Jisi from Peking University said there had been "an unprecedented number of heated debates" among Chinese analysts in which some optimists predicted that China would catch up with the US economically in 20 years, while others disputed what they saw as "exaggerations of China's strength". The conference process had some success in producing consensus. Kurt Campbell reported "near universal acknowledgement" among participants that for the foreseeable future China's leaders would be "profoundly focussed" on domestic goals. With regard to US–China relations, however, participants envisaged two potentially diverging paths – one of relative stability and coexistence, and another characterised by "a host of frictions".[30]

Also in the mid-2000s, two US think tanks – the Center for Strategic and International Studies and the Peterson Institute for International Economics – commenced a significant new project that appeared initially to offer a promising, large-scale, ongoing, and institutionally supported approach for assessing China's likely evolution and impact. They agreed, however, that this would be "no easy proposition". The project took the form of a three-year "China Balance Sheet". Its initial product was a book which aimed, ambitiously, to generate a "firm factual and analytical footing" as well as "one stop shopping" for those who wanted "to grasp, in a single place, both the basics and the complexities" of China. This demanding task was addressed by several prominent US China experts who aimed to provide a comprehensive and holistic view, assisted by an advisory committee of other high-level experts. The project would provide thorough and objective information as well as further analyses, publications, conferences, and a dedicated website. The book's authors aimed to provide "a badly needed foundation" for "comprehensive, authoritative, and accessible information and analysis".[31]

In the event, however, it seems fair to conclude that the project failed to fully live up to these commendable but exacting aims. It relied primarily on the views and judgements of individual experts, albeit highly qualified and experienced. They sought to address critical questions about China's future within four subject categories, namely *Continued growth or collapse?*; *Democratization or disorder?*; *Economic opportunity or threat?*; and *Security partner or rival?* Within this framework, the authors largely succeeded in maintaining a balanced approach. Dealing with the question of economic growth or collapse, Nicholas Lardy (one of the US's most-qualified and best-known China economists) noted its impressive performance but emphasised that straight-line extrapolations were "rarely warranted" and that China needed to transition to a new consumption-based growth path. The democratisation or disorder question was addressed by Bates Gill who perceived that China's leaders needed to balance pressure for political reform with concern for the CCP's survival. Addressing the question of whether China would present an opportunity or threat for the world economy, Lardy thought on balance it was an opportunity but it risked a protectionist reaction. Regarding whether China would be a security partner or rival, Derek Mitchell felt Beijing's stated priorities – to create a favourable international environment for its economic development and to preserve its territorial integrity – were indeed its primary motivations, although it could become more assertive.[32]

The China Balance Sheet Project continued to be active over its planned three-year period. A conference was held titled *The China Balance Sheet in 2007 and Beyond* with further papers by individual experts. A third and final publication appeared in 2008. A forceful and apprehensive appraisal by Fred Bergsten concluded that China had already become "a global economic superpower", with the potential to "disturb the existing international order and trigger security as well as economic conflicts". On the other hand, he thought a cooperative partnership between China and the US was "not only possible but also seemingly inevitable". Charles Freeman thought China was "far from being on the brink of collapse" and the CCP would continue in power "for the foreseeable future".[33]

An even larger endeavour to bring together many experts to forecast China's future occurred within the context of a series of major reports sponsored by the US National Intelligence Council. In 1997 the NIC published *Global Trends 2010*, the first of five lengthy reports analysing future global developments. Although these reports covered trends across all countries, China – as the largest emerging power and the biggest potential rival to the US – had a very high profile. The assessment process brought together experts from all sectors of government, research, and business to contribute to an agreed assessment of the global outlook, including many with China expertise. The tone of these five reports was generally positive about China's prospects, but they also emphasised there was considerable uncertainty.

Global Trends 2010 concluded that although China had considerable economic and military potential it was "beset by significant internal problems" and its power would grow only gradually.[34] *Global Trends 2015*, produced in 2000, saw China's future as "fraught with unknowables" and with pressures that would constrain its economic growth and threaten the regime's survival. However, it might not be overwhelmed by these problems. Either a "weak, disintegrating China" or an "assertive China" could pose a challenge to other countries, but a "more commonly held view among experts" was that China's foreign policy would continue to be peaceful. Nevertheless, the People's Liberation Army (PLA) could eventually become the region's most powerful military force.[35] The third report, *Mapping the Global Future* published in 2004, examined key global trends to 2020. Experts from around the world were consulted in a series of six regional conferences on four continents. The report sought to identify "Relative Certainties" (e.g. the rise of Asia) and "Key Uncertainties" (e.g. whether the rise of China/India would occur smoothly).[36]

Perceptions of China's status in world affairs received a further boost with the publication in 2008 of the NIC's fourth report *Global Trends 2025: A Transformed World*, which claimed to be "the most collaborative yet produced", and included a workshop in Beijing. Its methodology focused on key variables that it judged would have a disproportionate influence. It considered it a Relative Certainty that a global multipolar system was emerging, in which China was "poised to have more impact on the world over the next 20 years than any other country". The United States would be only "one of a number of important actors". But the report indicated that in 2025 US national power would still be over 50 percent higher than China's. It also predicted that China's growth "almost certainly will slow, if not recede". Each of China's problems would be soluble in isolation, but in combination they could produce "a perfect storm". The report noted its judgements could be upset by surprises and unintended impacts, in an era likely to contain "more contingencies than certainties". Finally, it emphasised that "leadership matters", because no trends were immutable and effective intervention by key actors could "decrease the likelihood and severity of negative developments and increase the likelihood of positive ones". Human actions would probably be "the critical determinants". Leaders and their ideas would be, as previously in history, "the biggest game-changers".[37]

As noted in Chapter 1, the NIC's fifth report *Global Trends 2030* introduced a new focus on these "game-changers". They were seen as crucial because the future was not "set in stone" but the result of interplay between "megatrends, game-changers and, above all, human agency". Several "Black Swans" were also listed including the possibility of an "economically collapsed China".[38] A paper by Michael Horowitz and Philip Tetlock about these *Global Trends* reports found that they placed too much emphasis on conventional wisdom and engaged in "extensive hedging". They accepted, however, that criticism of the reports was not entirely fair. A high level of accuracy could not be expected when predicting the long-term future due to the prevalence of "irreducible uncertainties" that made forecasting "something that falls between difficult and impossible". Nevertheless they thought such reports were still worth the effort given the potentially huge benefits that might be obtained from even small improvements in the accuracy of probability judgements.[39] These comments are a fitting epitaph for the forecasting achievements of all the experts groups surveyed above. Overall, and especially compared with individual experts, they produced notably more successes than failures – but still fell short of what might perhaps have been accomplished with more persistent efforts.

Scenario construction

Among China watchers scenario construction has enjoyed enduring popularity as a forecasting method, and has consumed a great deal of analytical effort. A large part of the appeal of this strategy was that the use of scenarios explicitly acknowledged the existence of uncertainty. It explicitly assumed the possibility of two or more different outcomes, even though experts often thought some scenarios more likely than others. Critics might argue that this method was not necessarily effective in narrowing the range of uncertainty. This weakness led Robert Fogel, for example, to comment that: "When speculating about the future, the range of possible scenarios is virtually unlimited. The point at issue is not what might occur, but what is likely to occur."[40] A number of China experts took up this challenge and were quite courageous in venturing to predict which of their scenarios were most likely to eventuate. Others displayed impressive ingenuity in developing new and innovative scenario construction techniques.

Possible scenarios

Many analysts chose not to risk assigning probabilities to their scenarios and were inclined instead to abandon themselves to uncertainty. They were quite willing to accept the plausibility of both optimistic and pessimistic scenarios. To be fair to them, this reflected the usual approach to scenario construction. Orville Schell, for example, once advised businessmen planning their involvement with China to sketch out "with as much logic as you can" at least two possible outcomes – "the most dire scenario as well as the most hopeful scenario, and maybe something in between".[41]

The post-Tiananmen period spawned numerous scenarios that speculated about a wide range of future possibilities but offered only very limited evaluation of their credibility. Christopher Clarke offered five "mid-term possibilities" ranging from *Descent into Chaos* to *Continued "Dictatorship of the Proletariat"*.[42] Chu-yuan Cheng's scenarios ranged from *Hardline Dominance* to the *Rise of Regional Independence*.[43] Harry Harding considered scenarios based on the assumption of either successful or unsuccessful reform. He thought successful reform was likely but "by no means inevitable".[44] Revisiting these scenarios later, Harding dismissed the "most extreme outcomes", but believed it "equally likely" that China's future would be peaceful or that it would sink into internal decay.[45]

In the late 1990s Harvard's Martin Whyte commented that predictions about trends in China had yielded two popular but "wildly divergent" scenarios – a "stability scenario", and a "chaos scenario". He too thought both scenarios had credibility. The continuation of "muddling through" might produce a state of "stable unrest" but chaos was "by no means a remote possibility".[46] Stepping forward to the 2000s, when China's rise seemed more assured, Kenneth Lieberthal nevertheless also suggested a wide range of possible configurations, from a successful country with continued one-party dominance to a corrupt authoritarian system with slower growth and declining international stature. He also considered China's possible emergence as a weak democratic state or a nightmare scenario with a divided elite, a contracting economy, environmental decline, territorial losses, and a declining international position.[47] Analysts at McKinsey and Company listed three possible paths: *Anarchic China, Authoritarian China* and *Advancing China*, declining to indicate which they thought most likely.[48] Paul Monk proposed four possible "models" for China, none of which he considered were "predetermined". Only one of these was optimistic: namely, *Mutation*, in which the CCP's role would be reduced. He also considered *Maturation*, in which China kept "its head above water" but failed to achieve economic transformation; *Metastasis*, in which a repressive CCP opposed change; and *Militarisation*.[49]

A major effort to construct scenarios for China's future was undertaken by the US National Bureau of Asian Research. It aimed to develop alternative scenarios in 2020 for China's foreign policy, domestic politics, economy, and society. Within each of these four areas, a China expert was tasked to present alternative scenarios. There was a deliberate emphasis on possible rather than probable outcomes, on the basis that this should help to "avoid being surprised by the unexpected". Regarding foreign policy, David Lampton considered China's posture might be *broadly cooperative* or *at odds* with the international community, venturing only that a cooperative scenario was unlikely but a conflictual scenario was avoidable. Regarding domestic politics, Cheng Li considered the possible emergence of a *Democratic China, Prolonged chaos*, or *A resilient, authoritarian regime*. He favoured the first scenario, but accepted there could be "multiple possibilities". Regarding China's economy, Peter Bottelier foresaw different scenarios depending on the extent of progress with reforms, from *A successful second transition* to *Crisis management*. Fenggang Yang identified three ideological currents in contemporary Chinese culture. Selected elements from each of these four scenario constructions were then combined to present, allegedly, four "distinct, plausible, and internally coherent futures": namely, *Democracy with Chinese characteristics*, *Strong CCP faced with economic challenges*, *Inner-party democracy, international trouble*, and finally *Doomsday* characterised by declining central control, a "perfect storm" of economic failures, loss of legitimacy, and international contention, leading to China "spiralling down a cataclysmic trajectory". These four scenarios aimed to provide "a mechanism to bound uncertainty" and to "fine-tune and diversify" decision-making processes, although it seems reasonable to observe that this might prove difficult given that decision-makers were offered such a diversity of possible outcomes.[50]

Two other experts – Kerry Brown and Michael Santoro – subsequently used an approach that focused on opposing scenarios but again avoided clear endorsement of any. Brown considered *Stable China* and *Unstable China*, while Santoro considered *Pax China* and *Nox China*. Brown considered that elements of both scenarios were already present in China. Santoro chose to dismiss the likelihood of an intermediate *Muddling Through* scenario because he thought China's volatility would not support "benign gradualism".[51]

Probable scenarios

Not all experts were so reluctant to assign probabilities to their scenarios. Some accepted a responsibility to provide their best estimate of future developments. Michel Oksenberg and Kenneth Lieberthal, in the mid-1980s, went so far as to assign percentage probabilities to their five scenarios for the future of reform, which ranged all the way from *Successful Reform* to *Abandonment of Reform* and *Systemic Breakdown*. They thought *Abandonment of Reform* was unlikely (20–25 percent) and saw only a small chance (5 percent or under) of *Systemic Breakdown*. *Successful Reform* seemed most likely (35–40 percent), leading to "the genuine emergence of this huge nation as a major player on the world scene".[52] In the mid-1990s, Michael Swaine assessed the probability of discontinuities in China's development. He judged that *A Highly Assertive China* was unlikely, while *A Weak, Insecure, and Defensive China* that might eventually lead to "a complete collapse of the Chinese state" was also unlikely. A best-case scenario would be *A "Normal" China, Pursuing Much Greater Cooperation with the West*, which was also unlikely in the near- to medium-term.[53]

Another mid-1990s scenario analysis, which with the benefit of hindsight was quite successful, was conducted by Richard Baum. His aim was to sort out some "conflicting claims and predictions" by examining, very tentatively, the pros and cons of each of ten scenarios within a modest three to five year timeframe. Baum's ten scenarios included three "evolutionary" scenarios: *Muddling Through*, *Neo-conservatism*, and *Neo-authoritarianism*; and four "devolutionary" scenarios: *Immobilism*, *Regional Fragmentation*, *Chaos*, and *Praetorian Military Intervention*. The remaining three "revolutionary" scenarios included two variants of a *Democratization* scenario: *Breakthrough from Above* and *Revolution from Below*, as well as *Neo-Maoist Revival*. He assigned odds (from 3–1 to 50–1) for each scenario. In order of probability, he considered *Neo-conservatism*, *Neo-authoritarianism* and *Muddling Through* to be "more likely", as has indeed been reflected in subsequent developments. His "less likely" scenarios were *Immobilism*, *Praetorian Military Intervention*, *Democratic Revolution From Above* and *Regional Fragmentation*; and his "least likely" scenarios were *Democratic Revolution From Below*, *Chaos*, and *Neo-Maoist Revival*. He observed aptly that the *Chaos* scenario was "the hardest to forecast" because its onset would be "non-linear and sharply discontinuous", but thought in China there were many powerful mitigating factors against this scenario.[54]

A similar instance where China experts attempted to quantify scenario probabilities was revealed by Joseph Nye. With Ezra Vogel he produced six scenarios in 1994 for the US government on China's outlook over the following decade. They then asked "some academic specialists" to assign probabilities to them. The result was *East Asian–Style Development* (probability 40 percent), *Democratic Reform* (probability 10 percent), *Pinochet China* with authoritarian government and continued rapid growth (probability 10 percent), *Modern Middle Kingdom* marked by nationalism and xenophobia (probability 10 percent), *Federated Feudalism* (probability 20 percent), and finally *Stagnation* (probability 10 percent). Nye noted that the first three scenarios, with a cumulative probability of 60 percent, all assumed high economic growth and relatively moderate foreign policies.[55]

Another unusual and ambitious approach to scenario construction was attempted in the mid-2000s in a book edited by South Korea's Jae Ho Chung titled *Charting China's Future.* It aimed to identify "possible future contingencies for China's political, social and international dimensions". Each of the expert contributors was asked to assess the likelihood of several scenarios based on their "best informed guesses". Chung was commendably cautious about the project's limitations, given that all discussions of China's future must be "inconclusive at best". He suggested several scenarios: from "*the Yugoslavian model*" where China fragmented, to "*the coexister*" with China seeking a cooperative relationship with the US or "*China threat*" where a strong China mounted a revisionist challenge.[56]

Among the book's contributors, Bruce Dickson's most likely scenario for China's political future was continued CCP rule, Yawei Liu's favoured scenario for possible electoral reform was a "Putin solution" with limited democracy, and Tao-chiu Lam's most likely scenarios for possible federalism were either the status quo or reduced local autonomy. Chung himself discussed scenarios for social unrest, with "stable instability" considered the most likely. Shiping Tang's chosen scenario for China's foreign relations was "crumbling unipolarity". Peter Gries' most probable scenario for US–China relations – selecting from friends, partners, rivals, or enemies – was rivals. Finally, Jean-Pierre Cabestan's scenarios for the future of Taiwan rated "status quo but with no solution" as the most probable. Chung then took the novel approach of amalgamating all of the scenarios rated as most likely; resulting in a predicted outcome of continued CCP rule with limited democracy, relatively successful domestic conflict management, and China emerging as a limited global player but with tense relations with the US especially over Taiwan. But Chung emphasised these outcomes could be altered by "chance factors" or "unexpected determinants".[57]

It is of more than passing interest that, among those experts who attempted to assign probabilities to their scenarios, all were relatively successful. None could be said to have failed spectacularly. This limited sample therefore suggests, at least tentatively, that there are strong grounds to argue that scenario constructors should not feel unduly inhibited about attempting to calculate probabilities because this may significantly increase the usefulness of their efforts. The same success was not enjoyed, however, by another group of experts who were uniformly pessimistic about China's prospects and chose to concentrate their attention primarily on determining *which* pessimistic scenarios might eventuate, giving only limited recognition to the presence of uncertainty.

In the immediate post-Tiananmen period, for example, Jurgen Domes conceded the theoretical possibility that the ruling CCP elite might survive but favoured four more-pessimistic scenarios. Of these four, the two he thought more probable were, first, a long period of repression leading to "a large-scale, extremely violent confrontation" in which communist rule would "drown in a veritable ocean of blood", or, second, a deteriorating economy causing a leadership crisis and transition to another political system.[58] Geoffrey Murray, in the late-1990s also examined several pessimistic scenarios. He thought China might be "strong or weak, a superpower or an under-achiever". A strong China would not be "a house-trained pussy-cat" but "a veritable tiger". Its disruptive influence could be similar to or greater than the Kaiser's Germany or twentieth century Japan, and its pursuit of territorial claims could make it "a prickly presence" in Asia. A weak China scenario could result if its growth bubble burst and it disintegrated "into squabbling regional fiefdoms".[59]

While speculating about how China might escape the "transition trap" Minxin Pei also suggested several scenarios, none of which were encouraging. They included economic decline and rising social tensions leading to a Soviet-style anti-regime revolution; absence of reform leading directly to regime collapse; and regional divergence as different areas either prospered or deteriorated. He also did not rule out a prolonged period with the CCP still in power but the country "mired in misrule".[60] A further set of mostly negative scenarios was proffered by French author Guy Sorman, from rebellion breaking out "all over China", to declining growth or even bankruptcy. He personally advocated another scenario in which outside intervention would restore human rights and freedom of speech.[61]

Innovative scenarios

In addition to scenario constructions that declined to attach probabilities and those that did, others were distinguished by innovative and/or resource-intensive techniques. These included, for example, an elaborate set of scenarios devised by two leading US authorities on scenario planning, James Ogilvy and Peter Schwartz.[62] They approached their task "neither as futurists nor as experts on China" but relied instead on "a subtle blend of trend analysis and well-tutored imagination". They presented three "storytelling" scenarios, each describing a possible future in 2020. These, they said, were unlikely to unfold as presented but they would reveal "coherent sets of possibilities that lie imbedded in the present". They accepted that China was "a complex adaptive system with billions of variables", but whatever happened "some reflection of it" should be seen in the scenarios and you would therefore be "unlikely to get caught flat-footed".

Prior to setting out their own scenarios, Ogilvy and Schwartz commented on three popular conventional wisdom scenarios. In Scenario 1, *Singapore x 70*, China became "a disciplined, benign, tightly integrated economic powerhouse" just like Singapore only bigger. In Scenario 2, *Soviet Devolution*, China would break up, peacefully or violently, like the Soviet Union. In Scenario 3, *The Big Bully*, China would be "belligerent, brutal, harshly centralized, and absolutist" like North Korea. They thought China's future was likely to be more surprising than any of these conventional scenarios. They claimed they would go "into the smoke and thunder of the great forces driving China" and come up with non-conventional scenarios that were "equally plausible" but based on different logics, rather than on "a single model pushed faster or slower, higher or lower, optimistic or pessimistic, best case or worst case".

Whether Ogilvy and Schwartz's scenarios succeeded in meeting these lofty ambitions is extremely questionable. Although we have not yet reached 2020, their scenarios have a fanciful and elusive quality. Scenario One, *The China*

Web, imagined China becoming "a massively decentralized village-based economy". It would leapfrog the industrial revolution and build a relatively clean economy. In Scenario Two, *The New Mandarins*, kinship and connection (*guanxi*) would become "the matrix of a new China". Beijing would provide the framework for this "network of networks", but get "out of the business of regulating people's lives and thoughts". In Scenario Three, *The Thief of Beijing*, China "collapses into a morass of decadence, corruption and greed" with "100 million robber barons clawing for wealth at any cost" and the environment becoming "a sewer". Eventually a charismatic military leader emerges as a saviour figure and heads "the world's largest kleptocracy".

Another interesting approach was taken by the US-based Global Business Network (GBN), for a gathering of business representatives and invited speakers that addressed China's prospects in the decade to 2015. Orville Schell noted it was "between transitions" and could be either a rising superpower or a tinderbox. Participants compiled and prioritised numerous "critical uncertainties", which were used as the basis for scenario exercises.[63] These critical uncertainties included China's impact on the international economy, its global role and relationships, its military/strategic orientation, and its environmental challenges. Optimistic scenarios included *Panda to Dragon* with China emerging as a robust economic actor and *The United States of China* with multi-party elections. Pessimistic ones included *Empire Strikes Back* with a police state controlling social unrest and *China, the Black Hole* with mismanagement and domestic collapse. A second GBN scenario exercise asked how long China's "winning streak" could last given ongoing threats from "corruption, environmental crisis, and social turmoil". It presented four scenarios based on the popular quadrant method, with the horizontal axis representing a China with either low or high economic influence, and the vertical axis representing a China that resolved conflicts either diplomatically or militarily. This method resulted in four possible scenarios. For example, in a "dream scenario" labelled *Emperor of Business* China had high economic influence and resolved conflicts diplomatically.[64]

Another two China-focused scenario construction exercises also used this quadrant approach. The World Economic Forum (WEF) explored scenarios for the following 20 years during a series of workshops with high-level participants in Singapore, Washington, London, Shanghai and Beijing. The scenarios aimed to explore "the complex interplay of underlying drivers and critical uncertainties". The horizontal axis in WEF's quadrant indicated whether or not China would implement effective internal institutional reforms and the vertical axis whether or not the global environment would be inclusive or discriminative for China. The *New Silk Road* scenario, for example, considered the most favourable quadrant where the global environment was inclusive and reforms were effective. The report encouraged organisations to develop strategies and policies "that would work in all scenarios".[65] A similar quadrant was developed for the City of London by SAMI Consulting and Oxford Analytica, considering alternative visions for China (and India) to 2015. The two principal drivers selected were (on the vertical axis) whether or not state institutions and commerce were sufficiently adaptive and effective and (on the horizontal axis) whether the outside world saw China as a source of wealth and opportunity or as an economic threat and source of instability. (Both of the WEF and SAMI/Oxford scenario constructions therefore saw the two critical drivers as being effective or ineffective governance and a positive or negative external environment.)[66]

A further major scenario construction exercise was conducted by the Center for Global Affairs (CGA) in New York. It wanted to develop scenarios that could provide "penetrating insights" and identify "key actors, drivers, and impediments of change" as well as possible wild cards. It resulted in "three stylized future trajectories" for China in 2020, developed through "a facilitated conversation" with 18 China experts. The first of CGA's three chosen scenarios was *Fragmentation*, in which challenges to central authority placed the survival of the CCP at risk; in the second, *Strong State*, China remained highly autocratic but rapid growth continued; and in the third scenario, *Partial "Democracy"*, the CCP was forced by popular demands to adopt a "new brand of participatory authoritarianism".[67]

Robert Broadfoot and Michael Enright also employed an innovative and well-structured approach to scenario development. They identified four "key uncertainties", including China's ability to maintain rapid economic growth, its social cohesion, cross-Taiwan Straits relations, and relations with other great powers. For each of these uncertainties two possible outcomes ("good" or "poor") were considered, which when combined in a "Scenario Tree" resulted in a total of 16 possible scenario outcomes of which five were given detailed treatment. The authors stated they had not attempted to forecast China's future but only to increase understanding of the "parameters of uncertainty".[68]

It is difficult to fault in principle most of the scenario methods discussed above. They all had the virtue of raising for consideration a number of alternative futures that might otherwise not be considered. Collectively they raised awareness of many different possible outcomes. Some of these endeavours, with hindsight, stand up quite well. They certainly gave due consideration to the presence of uncertainty, and probably increased appreciation of the ability of human actors to influence or change trends and outcomes. These benefits were often not without drawbacks, however. Many scenario exercises could be criticised for being either poorly researched, naïve or fanciful, lacking in depth, or unhelpful to policy-makers in narrowing the range of future possibilities, as well as for failing to provide a useful mechanism to evaluate conflicting expert opinions.

Credibility of expert opinion

There is a continuing high level of public interest in the judgements and predictions of experts. Collectively, however, they have begun to acquire a questionable reputation due partly to greater awareness that poor judgement is frequently not held accountable, as exposed by Philip Tetlock. Dan Gardner's book *Future Babble: Why Expert Predictions are Next to Worthless, and You Can Do Better*, was greeted by one reviewer with the hope that it would "shock the human race, and particularly the chattering expert class, into a condition of humility about proclamations about the future".[69] There seems little room for doubt that, in the case of China, a major reason why expert credibility is in decline is because they have failed to deal effectively with the challenge of uncertainty.

The preceding analysis of forecasts confirms there are good grounds for doubts about the credibility and reliability of experts. China watchers predicted that its economy would collapse, that the CCP would disappear, that leadership successions would be unstable, that regionalism would lead to national disintegration, that social unrest would lead to revolution, that the financial system would fail, that the economy could not succeed without democracy and a free market, and that as it grew stronger China would seek to dominate Asia and challenge other major powers. Even optimists frequently underestimated China's economic potential and overestimated its international influence. It continues to be striking that the most extreme of the possible outcomes – internal collapse and external domination – have both featured so prominently in the literature and continue to have many adherents.

Sometimes, however, expert forecasts about China's future tended to contradict critics like Tetlock and Gardner. Despite their failures, China watchers can also point to some degree of predictive success, or at least partial success. Many certainly exemplified the virtues of Tetlock's foxes – "those humble types who accept the future is very difficult to predict". China watchers who had acquired expertise through long-term scholarly research and direct exposure to the country produced many useful analyses and insights, and often came quite close to predicting actual outcomes. These experts showed admirable humility and awareness of the difficulty of their task. They were often willing to revise their opinions. They resisted extremes of judgement and accepted the importance of balanced and sustained research.

It can be argued that some of the most prominent pessimists and optimists also played a useful role – the pessimists by emphasising the real risks and challenges China faced and continues to face, and the optimists by pointing to the possibility that China might surprise observers by doing better than expected. The evidence, in the case of many China experts, tends to challenge Tetlock's finding that greater expertise, or even fame, may imply declining reliability of judgement. In short, it would seem inappropriate to be too quick to dismiss the value of the forecasting efforts of China watchers. Rather, perhaps, they should be commended for tackling such a difficult task. That said, however, there is also a cautionary message for the all-too-numerous "hedgehogs" who tended to become overconfident in their predictive abilities and were less focused than they might have been on the need to avoid confirmation bias, to consider contrary evidence, and to acknowledge uncertainty.

It also deserves to be recognised that to their credit China watchers were often among the most constructive critics of their own field. They frequently acknowledged the problems that confronted them, and their reflections and advice contained much insight. Cheng Li, for example, commented that inaccurate or false predictions by Western analysts highlighted "the deficiencies and inadequacies in the field of China studies". A common problem among researchers, he believed, was that they tended to use "unverified sources, conventional approaches, and old analytical frameworks" as well as basing their arguments on "rumours, speculations, and myths".[70] Aaron Friedberg believed experts lacked the kinds of "powerful predictive tools" needed to forecast China's future. Nevertheless, he urged them to attempt to "find ways of apprehending the full array of causal forces".[71] He also accepted that forecasting China–US relations was "an inexact science", since politics was "simply too complex and contingent". It was easy to underestimate "the impact of personalities, unforeseeable errors, and random occurrences".[72]

Several experts have been especially critical of the persistent duelling between proponents of China's dominance or collapse.[73] Barry Naughton and Dali Yang expressed their perplexity about the inability of one group of experts to appreciate "the extent to which its own conclusions should be modified by the insights of the other perspective".[74] Josh Kerbel lamented that US predictions tended to range from "rising China" to "doomed China", begging the question about how individual predictors could feel "confident enough to make firm single-outcome predictions".[75] John Hoffman noted that hardly a day went by without a new publication about the "China miracle" or the "China threat". China was seen as either "on the verge of imminent collapse" or "on the brink of economic or political dominance". He marvelled that any nation could be the source of "so much contradictory information and confusing conclusions".[76] David Lampton argued that opinions about China were a reflection of observers' own subjective concerns. China was "a giant screen upon which outsiders project their hopes and fears".[77] Cheng Li also took up this theme. He advised China watchers to avoid cleaving "to conventional, old-fashioned perceptions of this rapidly changing country, taking special care to steer clear of dogmatic cynicism on the one hand, and ill-grounded optimism or wishful thinking on the other".[78]

Those China experts who acknowledged the high level of uncertainty surrounding their subject emphasised that their forecasts could be upset by unforeseen surprises or disproved by the tendency for important actors, especially China's leadership, to take corrective actions. Andrew Nathan commented that analysing China's "current trajectory" was "not the same as predicting what will happen in the future", because unexpected events could throw any country off track. This had often happened in China "and probably will again".[79] Others, like Chetan Ahya maintained that China was differentiated from other emerging countries by the "single-minded determination of the politicians to 'make it happen' ".[80] Barry Naughton, looking back over several decades, discerned a consistent pattern that "when problems in China get really, really dramatic so that policy-makers are forced to address them, then they do". He added that this might only occur when the government was "pushed hard enough so that they have no choice but to tackle the problem", but when this point was reached their record was "extremely good".[81]

Nevertheless, although many China watchers were clearly aware of their forecasting failings and of the reasons for them, they seldom offered constructive suggestions as to how they might be addressed holistically. The following chapters exemplify an alternative and complementary approach – based on an expanded use of the concept of critical uncertainties – that may point to a more effective way to structure and evaluate the expert debate on the most critical and consequential issues in China's future.

Notes

1 Avery Goldstein, "Political Implications of a Slowdown", *Orbis* 43, no. 2 (Spring 1999), 203–221. Goldstein's caution was appropriate because his own forecast about China's economic prospects in the new century also proved inaccurate. He specifically ruled out a "much better than expected" scenario in which China would return quickly to double-digit growth.

2 Orville Schell, *China: Boom or Boomerang* (Truthdig.com, 3 December 2005).

3 Orville Schell, *China's Contradictions – and Possible Collapse* (Global Business Network, September 2001).

4 Orville Schell, *Red Star Rising/China – The Coming Collapse* (RedHerring.com, 6 November 2002).

5 Kenneth Lieberthal, "China in 2033: What will China Look Like in 30 Years?", *China Business Review* (March/April 2003), 42–48.

6 Willy Wo-Lap Lam, *Chinese Politics in the Hu Jintao Era: New Leaders, New Challenges* (Armonk, NY: M. E. Sharpe, 2006), xiv, 32, 247–254 and 284–285.

7 Nathan, "Present at the Stagnation", 181–182.

8 David M. Lampton, "Paradigm Lost: The Demise of 'Weak China'", *The National Interest* (Fall 2005), 73–75.

9 James Kynge, *China Shakes the World: A Titan's Rise and Troubled Future – and the Challenge for America* (Boston, MA: Houghton Mifflin, 2006), xi-xvi, 34–53, 127, 154–155, 183–187 and 233–240.

10 Joseph Fewsmith, ed. *China Today, China Tomorrow: Domestic Politics, Economy, and Society* (Lanham, MD: Rowman & Littlefield, 2010), 278–283.

11 George Walden, *China: A Wolf in the World?* (London: Gibson Square, 2008), 18–19, 244–246 and 271.

12 Kerry Brown, *Struggling Giant: China in the 21st Century* (London: Anthem, 2007), 139–141.

13 Kerry Brown, "How China is Weaker Than it Looks", *The Diplomat* (12 October 2010). See also Brown's "China Will Get Democracy", *The Diplomat* (8 April 2010) and *China 2020: The Next Decade for the People's Republic of China* (Cambridge, UK: Woodhead, 2011), xviii and 19–43.

14 "Debating China's Future", *China Security* 4, nos. 2 and 3 (2008), 2–26 and 3–31 respectively.

15 David Shambaugh, "China's Immediate Future: Stable or Unstable?", in *Charting China's Future: Domestic and International challenges*, ed. David Shambaugh (London: Routledge, 2011), 173–178.

16 Charles Wolf Jnr *et al.*, *Fault Lines in China's Economic Terrain* (RAND National Defense Research Institute, 2003), xv.

17 Ibid., xvi-xxi, 6 and 169. Also questionable is why the study did not address other, and possibly more important, potential fault lines. It acknowledged that its coverage was "selective, rather than exhaustive".

18 June Teufel Dreyer, "The Limits to China's Growth", *Orbis* 48, no. 2 (Spring 2004), 233–246.

19 Jonathan Laing, "What Could Go Wrong With China?", *Barron's* (31 July 2006).

20 Susan L. Shirk, *China, Fragile Superpower: How China's Internal Politics Could*

Derail Its Peaceful Rise (Oxford, UK: Oxford University Press, 2007), 1–7, 20–64 and 255.

21 Wing Thye Woo, *What are the High-probability Challenges to High Growth in China?* (Brookings Institution, 2007).

22 Wing Thye Woo, "Assessing China's capability to manage the high-probability risks to economic growth: fiscal, governance and ecological problems", in Keun Lee, Joon-Han Kim, and Wing Thye Woo, eds. *Power and Sustainability of the Chinese State* (London: Routledge, 2009), 92–94.

23 Nicholas Eberstadt, "Will China (Continue To) Rise?", in *The Rise of China: Essays on the Future Competition*, ed. Gary J. Schmitt (New York: Encounter Books, 2009), 131–153.

24 David M. Lampton, "What if China Fails?: We'd Better Hope it Doesn't", *The Wilson Quarterly* 34, no. 4 (Autumn 2010), 61–66.

25 One of the main exceptions was where large numbers of analysts contributed to the development of quantitative models to project economic and environmental trends, as will be discussed in Chapters 5 and 6. Studies by international organisations and by commercial and government agencies also typically combined the expertise of many analysts, although the process was often not transparent.

26 Vaclav Smil, "China's Future: A Delphi forecast", *Futures* (December 1977), 474–489.

27 Michael D. Swaine, *China: Domestic Change and Foreign Policy* (RAND, 1995), xiii–xiv.

28 Oksenberg, Swaine, and Lynch, *The Chinese Future*, 1–19.

29 Conference papers were revised and published in David Shambaugh, ed. *Is China Unstable? Assessing the Factors* (Armonk, NY: M. E. Sharpe, 2000).

30 Kurt M. Campbell and Willow Darsie, eds. *China's March on the 21st Century* (The Aspen Institute, 2007), 14–34, 86, 131–142 and 159–161.

31 C. Fred Bergsten *et al.*, *China: The Balance Sheet – What the World Needs to Know Now About the Emerging Superpower* (New York: Center for Strategic and International Studies and Institute for International Economics, 2006), ix–xii and 2.

32 Ibid., 2–3 and 18–154.

33 C. Fred Bergsten, Charles Freeman, Nicholas R. Lardy, and Derek J. Mitchell, eds. *China's Rise: Challenges and Opportunities* (Peterson Institute for International Economics and Center for Strategic and International Studies, 2008), 9–31, 57–70 and 105–130.

34 *Global Trends 2010*, (United States National Intelligence Council, November 1997).

35 *Global Trends 2015: A Dialogue About the Future With Nongovernment Experts*, (United States National Intelligence Council, December 2000), 16–17 and 53–54.

36 *Mapping the Global Future: Report of the National Intelligence Council's 2020 Project*, (United States National Intelligence Council, December 2004), Foreword, 40 and 117.

37 *Global Trends 2025: A Transformed World*, (United States National Intelligence Council, November 2008), Foreword, iv–xii, 1–14, 29–30, 87 and 98–99.

38 *Global Trends 2030: Alternative Worlds*, (United States National Intelligence Council, December 2012), Foreword and ii–xi.

39 Michael C. Horowitz and Philip Tetlock, "Trending Upward: How the Intelligence Community can better see into the Future", *Foreign Policy* (6 September 2012).

40 Robert W. Fogel, *Why China is Likely to Achieve its Growth Objectives* (National Bureau of Economic Research, 2006), 12.

41 Schell, *China's Contradictions*.

42 *China's Economic Dilemmas in the 1990s*, 8–13.

43 Chu-yuan Cheng, *Behind the Tiananmen Massacre: Social, Political and Economic Ferment in China* (Boulder, CO: Westview, 1990).

44 *China's Economic Dilemmas in the 1990s*, 85–88.

45 Harding, "On the Four Great Relationships: The Prospects for China", 39–41.
46 Shambaugh, *Is China Unstable? Assessing the Factors*, 143–161.
47 Lieberthal, "China in 2033", 47–48.
48 Campbell and Darsie, *China's March on the 21st Century*, 25–34.
49 Monk, *Thunder from the Silent Zone*, 60.
50 Kuo *et al.*, "China in the Year 2020".
51 Brown, Struggling Giant, xiv–xvi and 2–8; Michael A. Santoro, *China 2020: How Western Business Can – and Should – Influence Social and Political Change in the Coming Decade* (Ithaca, NY: Cornell University Press, 2009), 132–137.
52 Michel C. Oksenberg and Kenneth Lieberthal, "Forecasting China's Future", *The National Interest* (Fall 1986).
53 Swaine, *China: Domestic Change and Foreign Policy*, 98–111.
54 Richard Baum, "China After Deng: Ten Scenarios in Search of Reality", *China Quarterly* 145 (March 1996), 153–175.
55 Joseph S. Nye, "China's Re-emergence and the Future of the Asia–Pacific", *Survival* 39, no. 4 (Winter 1997–98), 66–71.
56 Jae Ho Chung, ed. *Charting China's Future: Political, Social and International Dimensions* (Lanham, MD: Rowman & Littlefield, 2006), xi–xii and 1–5.
57 Ibid., 11–13, 165–190 and 203–204. A concluding chapter by Gilbert Rozman considered three further over-arching scenarios – successful economic growth, possible stagnation, and new reforms – with the first considered most likely for the coming decade, the second for the following decade, and the third for the following two decades.
58 Hicks, *The Broken Mirror*, 469–471.
59 Geoffrey Murray, *China, the Next Superpower: Dilemmas in Change and Continuity* (New York: St Martins, 1998), 1 and 209–217.
60 Pei, *China's Trapped Transition*, 210–212.
61 Guy Sorman, *Empire of Lies: The Truth About China in the Twenty-First Century* (New York: Encounter, 2008), 238–242. The book was first published in French in 2006.
62 James Ogilvy and Peter Schwartz, *China's Futures: Scenarios for the Fastest Growing Economy, Ecology and Society* (San Francisco, CA: John Wiley and Sons, 2001). Schwartz was the author of one of the then best known introductions to scenario planning: Schwartz, *The Art of the Long View: Planning for the Future in an Uncertain World.*
63 *China's Choices: A GBN WorldView Meeting Report*, (GBN Global Business Network, 2005). GBN noted that this meeting had been preceded by similar exercises over more than a decade.
64 Doug Randall and Jesse Goldhammer, "Four Futures for China Inc", *Business 2.0* (August 2006).
65 World Scenario Series, *China and the World: Scenarios to 2025*, (World Economic Forum, 2006), Preface, 7–9 and 66–67.
66 SAMI Consulting and Oxford Analytica, *Scenarios for India and China 2015: Implications for the City of London*, (October 2006), 1–11, 42–45 and 54.
67 Center for Global Affairs New York University, "China 2020", *CGA Scenarios* 3 (Winter 2009), 2–41.
68 Hoffmann and Enright, *China into the Future*, 267–311.
69 Dan Gardner, *Future Babble: Why Expert Predictions Are Next to Worthless, and You Can Do Better* (New York: Dutton, 2011). See also Dan Gardner and Philip Tetlock, "Overcoming our Aversion to Acknowledging our Ignorance", *Cato Unbound* (11 July 2011), and Louis Menard, "Everybody's an Expert: Putting Predictions to the Test", *The New Yorker* (5 December 2005).
70 Cheng Li, "The China Paradox and American Misperceptions", in *The United States and China: Mutual Public Perceptions*, ed. Douglas G. Spelman (Washington, DC:

Woodrow Wilson International Center for Scholars, 2011), 31. Zhang Qingmin suspected that shared analytical approaches among Western analysts led to "identical methods and views": see Zhang Qingmin, "Book Review: Jae Ho Chung, Charting China's Future", *China Review International* 14, no. 2 (Fall 2007).

71 Friedberg, "The Future of US–China Relations?"

72 Friedberg, *A Contest for Supremacy*, 33–37.

73 These competing views evoke Katzenstein and Okawara's interesting contention that it was becoming difficult to counter the view that scholarship was about "paradigmatic clashes", rather than "the disciplined analysis of empirical puzzles". See Peter J. Katzenstein and Noburo Okawara, "Japan, Asian-Pacific Security, and the Case for Analytical Eclecticism", *International Security* 26, no. 3 (Winter 2001/02), 154.

74 Naughton and Yang, *Holding China Together*, 2–24.

75 Josh Kerbel, "Thinking Straight: Cognitive Bias in the US Debate About China", *Studies in Intelligence* 48, no. 3 (2004).

76 Hoffmann and Enright, *China into the Future*, 1.

77 Lampton, "China: Think Again".

78 Li, "The China Paradox and American Misperceptions", 27–36.

79 Nathan, *Is Communist Party Rule Sustainable in China?*

80 Ahya *et al.*, *India and China: New Tigers of Asia.*

81 *Reframing China Policy Debate 2.* (Carnegie Endowment for International Peace, 1 December 2006).

Bibliography

Bergsten, C. Fred, Bates, Gill, Nicholas R. Lardy and Derek Mitchell. *China: The Balance Sheet – What the World Needs to Know Now About the Emerging Superpower.* New York: Center for Strategic and International Affairs and Institute for International Economics, Public Affairs, 2006.

Bergsten, C. Fred, Charles Freeman, Nicholas R. Lardy and Derek J. Mitchell, eds. *China's Rise: Challenges and Opportunities.* Peterson Institute for International Economics and Center for Strategic and International Studies, 2008.

Brown, Kerry, ed. *China 2020: The Next Decade for the People's Republic of China.* Cambridge, UK: Woodhead, 2011.

Brown, Kerry. *Struggling Giant: China in the 21st Century.* London: Anthem, 2007.

Chung, Jae Ho, ed. *Charting China's Future: Political, Social and International Dimensions.* Lanham, MD: Rowman & Littlefield, 2006.

Fewsmith, Joseph, ed. *China Today, China Tomorrow: Domestic Politics, Economy, and Society.* Lanham, MD: Rowman & Littlefield, 2010.

Gardner, Dan. *Future Babble: Why Expert Predictions are Next to Worthless, and You Can Do Better.* New York: Dutton, 2011.

Hoffmann, W. John and Michael J. Enright, eds. *China into the Future: Making Sense of the World's Most Dynamic Economy.* Singapore: John Wiley & Sons, 2008.

Kynge, James. *China Shakes the World: A Titan's Rise and Troubled Future – and the Challenge for America.* Boston, MA: Houghton Mifflin, 2006.

Ogilvy, James and Peter Schwartz. *China's Futures: Scenarios for the Fastest Growing Economy, Ecology and Society.* San Francisco, CA: John Wiley & Sons, 2001.

Shambaugh, David, ed. *Is China Unstable? Assessing the Factors.* Armonk, NY: M. E. Sharpe, 2000.

Shirk, Susan L. *China, Fragile Superpower: How China's Internal Politics Could Derail Its Peaceful Rise.* Oxford, UK: Oxford University Press, 2007.

Walden, George. *China: A Wolf in the World?* London: Gibson Square, 2008.

Wolf, Charles, Jr, K. C. Yeh, Benjamin Zycher, Nicholas Eberstadt and Sung-Ho Lee. *Fault Lines in China's Economic Terrain*. RAND National Defense Research Institute, 2003.

Woo, Wing Thye. *What Are the High-Probability Challenges to High Growth in China?* Brookings Institution, 2007.

4 Political uncertainties

The pervasive impact of uncertainty on attempts to forecast China's future suggests strongly that there is likely to be considerable value in a fresh approach that compares and evaluates the full spectrum of opinion among China experts about major issues within particular domains. This avoids placing undue reliance on the judgements of individual experts. It offers a way to move beyond a confusing and unproductive optimist–pessimist debate. It encourages supporters of both optimism and pessimism to recognise the realistic possibility of outcomes that fall outside the relatively narrow range they have previously been willing to entertain. It also increases awareness that the future is not yet determined and that it may even be channelled in new and positive directions.

The term *critical uncertainties* has been used by several experts, as noted above and especially in the context of scenario construction. But I use it here in a different and broader context. The concept of critical uncertainties offers, in my opinion, an especially valuable and productive analytical tool. It implies that there are a number of major issues affecting China's future about which it is not possible to offer definitive predictions and about which there is, consequently, a legitimate and ongoing debate. It therefore seems persuasive that an open-minded and comprehensive analysis and synthesis of these different expert views will be likely to yield a more robust understanding of the realistic range of future possibilities, and even perhaps suggest some of the more likely outcomes.

I recommend that the first step in this process should be to attempt to identify and specify the most important critical uncertainties. These can then serve as the starting point for an enduring but flexibly defined and continually redefined analytical framework. As indicated previously, these uncertainties are defined as "critical" because they are specified at a strategic or system level, and because they will have very high national, and usually also international, impacts and consequences. A well-defined structure of critical uncertainties can be used as a highly effective template to analyse the ongoing discourse among analysts about China's future and to provide a valuable foundation for further research. The end result of such a synthesis of analytical discourse will be a more comprehensive, nuanced, and balanced overview of possible developments. The wider employment of such a structured approach would extract greater value from existing research and could encourage more comprehensive and better-coordinated institutional approaches to the study of national futures, not only for China but also potentially for other countries.

As also suggested earlier, such critical uncertainties are probably best presented as a hierarchy of domain-specific interrogative questions, which will not endure for all time but will require regular updating and refining. This iterative process will inevitably be to some extent subjective and intuitive. For practical reasons, this study considers only four over-arching domains – political, economic, environmental, and international – and each of these domains is interpreted broadly and inclusively. For example, the political domain is interpreted to include the social underpinnings of the political system, the economic domain includes demographic aspects, the environmental domain includes natural resource and energy aspects, and the international domain includes military and security aspects.

In each of these domains, the overarching question is likely to focus on how effectively China will perform in that domain. In the political domain, the highest-level critical uncertainty can be expressed provisionally as: "*What are the prospects for China's political governance?*"; in the economic domain as: "*What are the prospects for China's economy?*"; in the environmental domain as: "*What is the outlook for China's environmental management and sustainability?*"; and in the international domain as: "*How will China's international relationships evolve?*".

This chapter will attempt to explore a second level of critical uncertainties within the political domain, based primarily on a review of the debate among China experts about these uncertainties. (Similarly, second level critical uncertainties for the remaining domains will be considered in the following three chapters.) It is worth noting at the outset that, when the prospects for political governance in China are discussed by experts outside China, their primary preoccupation is usually not – as might perhaps be expected – with the likely future *effectiveness* of the political system in delivering such public goods as prosperity, stability, and equity. Rather than focusing on effectiveness, a more compelling preoccupation among China watchers has been with the *nature* or *form* of the evolving political system, and especially with whether the current authoritarian system is capable of continuing to perform effectively or whether an alternative and more liberal political system is likely to evolve to better cope with future challenges and expectations.

In short, the focus of many if not most China experts has been on the institutional nature of the future political system rather than its deliverables. There has been, especially among analysts in the United States, a huge interest in the question: "*What are the prospects for continued authoritarian governance in China?*" and especially in the obverse of this question: "*What are the prospects for democracy in China?*" Two other related issues have also received much attention, namely "*Will China's political stability be threatened by social unrest?*" and "*What are the prospects for political liberalisation within China's authoritarian system?*" Given the preoccupation of China experts with these issues, and their intrinsic importance, this chapter will concentrate on the spectrum of views about these four critical uncertainties.

Authoritarianism

China's authoritarian regime has for many years given every external appearance of being well established and in control. There has nevertheless been a vigorous exchange of views among experts about the regime's durability. As foreshadowed in Chapter 2, a number of prominent China watchers have been very doubtful about the future of the CCP, including Minxin Pei and Will Hutton. Revisiting the issue in 2012, Pei concluded that following China's alleged entry into the "democratic transition zone" the durability of one-party rule was "doomed". Unless it "immediately" pursued "de-totalitarianism" it might find that a Soviet-style collapse was "its only future".[1] Hutton's emphasis was on China's inevitable political and economic failure if it did not adopt Western "Enlightenment Institutions" including a pluralist representative democracy.[2] Zheng Yongnian was also sceptical about China's authoritarian future. The Party, he said, was entering "uncharted waters" and if it faltered this might even lead to the breakup of the nation.[3] He described the future of the CCP as undoubtedly the key uncertainty for China. In his view "rigid flexibility" best described the CCP's strategy and such rigidity would be unsustainable "in the long run".[4]

Several other China watchers also stressed the multiple challenges and weaknesses confronting the authoritarian political system. The majority, however, appeared convinced that the CCP would nevertheless remain in power for many years. This view appeared to strengthen during the 2000s. Harvey Nelson found seven reasons why the CCP would endure, at least in the near to medium term, including that it could portray itself as the protector of stability, there were no likely alternatives, and it could manipulate nationalism. Moreover, China had no desire to follow the example of the Soviet Union so change would be approached "cautiously and incrementally".[5]

In an influential article in 2003 Andrew Nathan observed that after Tiananmen many observers, including himself, expected that the CCP would fall to a "third wave" of democratisation as suggested by Samuel Huntington. But the Party had reconsolidated. Regime theory held that authoritarian systems were inherently fragile due to weak legitimacy, over-reliance on coercion, over-centralisation of decision-making, and emphasis on personal power. But Nathan noted that the CCP's version of authoritarianism had proved to be "resilient", due in part to its success in institutionalising leadership succession, encouraging meritocratic promotion, and widening political participation. This weakened "too-hasty arguments that it cannot adapt and survive". Further explanations for the regime's success included the fact that living standards had risen, the CCP had co-opted elites, the population feared disorder, and perceptions of the CCP's unshakeable grip on power. He thought the available evidence proved the CCP was "not supine, weak, or bereft of policy options". China's leaders thought they could solve China's problems and they intended to address them. Arguments about the superior merits of democracy held no appeal for them.[6]

Nathan noted that scholars like Talcott Parsons, Richard Lowenthal, and Chalmers Johnston had debated the adaptability of totalitarian regimes for half a century, but the CCP had adapted significantly by abandoning utopian ideology and charismatic leadership, empowering a technocratic elite, reforming its bureaucracy, and reducing control over private speech and action – all without leading to regime change or to the democratic transition that Parsons predicted as the consequence of such adaptation. Nathan believed democratic transition in China was therefore not inevitable. China had already transitioned successfully from totalitarianism to authoritarianism, suggesting "an even more disturbing possibility: that authoritarianism is a viable regime form even under conditions of advanced modernization and integration with the global economy".[7]

In a later debate with Roderick MacFarquhar, Nathan continued to argue that current forces on balance supported the durability of the CCP's rule, rather than collapse or democratisation. A single spark, he argued, would *not* start a prairie fire due to the effectiveness of the CCP's fire brigade. The crucial factor in the regime's resilience was "the elite's will to power". He doubted that gradual modernisation of the CCP's rule would lead it to "grow out of authoritarianism". Changes to date had seemed "to consolidate rather than weaken authoritarianism". Nathan also maintained that three theoretical perspectives about China's future were being misapplied, his response being: "It ain't necessarily so". First, he argued that Seymour Lipset's theory that a rising middle class would demand democracy did not apply because China's middle class did not support democratisation. Second, Talcott Parsons' theory that a modern economy must function freely in order to operate efficiently did not apply because China's market-state economy so far had worked effectively. Third, the theory associated with Francis Fukuyama and others that socialisation in a global culture would produce democratisation did not apply to China because it wanted to develop its own non-Western alternative modernity.[8]

Such views were echoed and even amplified by Kjeld Brodsgaard and others, who argued that, instead of losing its grip the CCP was reasserting and strengthening its control. They contested the transition paradigm espoused by Thomas Carothers, which implied that the CCP would eventually wither away as China moved toward pluralism and democracy. Like Nathan, they found that instead of regime breakdown the evidence was for regime resilience and revitalisation.[9] Similarly, Jean-Pierre Cabestan emphasised the CCP's successful adaptation and use of "proto-democratic" institutions to support its legitimacy, which he believed supported a "flexible authoritarianism" rather than the "fragmented authoritarianism" perceived by others. He saw progress toward a CCP that he variously described as "consultative but elitist", "softer but stabilised", and "enlightened but plutocratic".[10]

Bruce Dickson initially appeared equally as impressed by the scale of the challenges facing the Party as by its newfound durability and adaptability. At the same time that he perceived an "unprecedented degree of institutional stability", he also saw a weakening of the CCP's strength at grassroots level. He questioned whether the CCP's co-option and recruitment of new elites would strengthen or weaken its hold on power.[11] His later research, however, led him to favour the view that the CCP's power would strengthen. He used the term "populist authoritarianism" to describe China's political system, noting that although conventional wisdom characterised it as beset by tension and turmoil, the CCP had been "more adaptable and more resilient than it often gets credit for". He saw China's problems as "more chronic than acute" and, unless there was an "unexpected flare-up", did not anticipate an imminent threat to the regime. The CCP benefited from a widespread belief internally that democracy was "not appropriate for China".[12]

In the first of three books about China's "red capitalists", Dickson argued that these entrepreneurs might "prop up an authoritarian regime" rather than undermine it. A "linear and deterministic approach" to the theory of political development, he asserted, ignored "the weak link between economic development and democracy". Adam Przeworski and Fernando Limongi, he said, had discovered that there was "no simple correspondence between economic change and the timing of democratization". The CCP might "remain in power indefinitely", even though lasting solutions might require additional political reforms.[13] Dickson's two following books indicated an apparent strengthening of his views about the essentially conservative and pro-regime attitudes of the red capitalists, who he saw as crony capitalists, allies of the state, and partners in the status quo.[14]

Although James Mann's view of the CCP's record on human rights and other issues was distinctly negative, he too saw little prospect that it would fade from the scene. He was strongly critical of a popular view in the West that China's economic progress and engagement with other economies would gradually lead to democracy. He dismissed this naïve optimism as the *Soothing Scenario*. He was also unconvinced by another view that China's weaknesses would lead to its collapse, which he called the *Upheaval Scenario*. Instead he thought it much more likely that the world would need to accept a more realistic *Third Scenario* in which the CCP would probably continue to be around for a long time.[15]

A major study by David Shambaugh pointed to evidence of both "atrophy and adaptation" in the CCP's development. He believed the Party had been "very proactive" in instituting political reforms, which belied the image popular among many analysts that China remained "an ossified Leninist state". Western analysts considered China's political reforms to be invalid unless they were "protodemocratic", but he believed the CCP's progressive adaptation had strengthened its legitimacy. At the same time as this adaptation was taking place, he also observed a progressive process of atrophy in the CCP's traditional instruments of control over the intellectual, social, economic, and political life of China's citizens. He judged that continuation of this process would also consolidate the Party's power. He disagreed with those who thought the party-state was in "a kind of free-fall". He concluded that it was adapting "fairly (but not entirely) effectively". However, he agreed it would be naïve to presume the CCP would remain in power perpetually and saw this as "an open question".[16] As China's political system evolved, Shambaugh's views moved more strongly in this sceptical direction. He considered that since 2009 the CCP had become more repressive and insecure. A coalition of vested interests had aligned with hard-line elements to pull back reforms. As it celebrated its 90th birthday in 2011 the CCP seemed "frozen in fear of the future". Like many 90-year-olds, the Party had become "increasingly infirm, fearful, experimenting with ways to prolong life, but overwhelmed by the complexities of managing it".[17]

In the new circumstances of the late 2000s, Andrew Nathan also appeared somewhat more inclined to emphasise the potential impermanence of China's authoritarianism than its resilience. On the one hand he was impressed by the CCP's vigorous longevity. It had fostered rather than stifled change, it had promoted competent leaders, ensured orderly succession, generated effective policies, and attracted popular support. But, on the other hand, he thought the Party suffered from the "birth defect" that its authority had never been subject to popular review as occurred in democratic systems. Authoritarian regimes, therefore, were "not forever" and they lived "under the shadow of the future". The CCP could only stave off crises by performing a flawless balancing act, and "it cannot afford to slip".[18]

Subsequently, however, three major studies of the CCP all concluded that it remained a viable and formidable organisation, provided it continued to reform politically and perform economically. Richard McGregor noted that the Chinese communist system was "in many ways, rotten, costly, corrupt and often dysfunctional", but it had proved to be "flexible and protean enough to absorb everything that has been thrown at it". Through a combination of co-option and coercion – and by exploiting both economic growth and nationalism – it had strengthened its position. He concluded that for the foreseeable future the CCP's leaders would "bestride the world as a colossus on their own implacable terms".[19] He added later that the idea of China becoming a democracy was "always a Western notion" born of theories about how political systems evolved. So far the evidence in the case of China suggested these theories were wrong. The Party had made it clear it did not want China to become a Western democracy, and had the means to ensure it did not.[20]

The second study, by Zheng Yongnian, agreed that the CCP was proactive and would dominate as long as it continued to transform itself. He interpreted the CCP's identity within a Chinese cultural context – as a transformed emperorship that accommodated some democratic elements. He believed that in the long haul democratisation was the CCP's only option, but not along Western lines.[21] The third study, by Kerry Brown, also argued that if the CCP continued to open up the political system it might still be around in half a century. He was not entirely convinced it would be successful, however, entertaining a "nagging suspicion" that the CCP's ambitions might in some ways be "fundamentally unsustainable". A "nightmarish combination" of economic, social, and environmental problems might also bring it down.[22]

Several other experts were ambiguous or uncertain about the CCP's longevity. Singapore academics John Wong and Lai Hongyi conceded that outside analysts had "constantly underestimated" the CCP's ability to adapt and to manage crises, and that the CCP's position was "not insecure". They also believed, however, that, without electoral democracy, fundamental flaws in China's political system could threaten the social order, especially if economic growth were to falter.[23] Willy Lam was impressed by the Hu Jintao–Wen Jiabao leadership's approach to socio-economic issues but concerned about lack of progress in reforming the "antediluvian" one-party system.[24]

Eric Anderson was more confident about the CCP's prospects. He doubted China's government would "significantly alter in form or function" in the next ten years, declaring confidently that: "A pragmatic, resilient, authoritarian regime will still be sitting in Beijing in 2020". However, he noted that unforeseen developments had "a way of derailing even the best laid plans".[25] Stephen Platt and Jeffrey Wasserstrom speculated that unless the CCP reinvented itself with "a new legitimizing narrative" or "a new multiparty political structure" it could risk being overthrown, but they warned that this might take "many generations longer" than expected.[26]

Social unrest

Many experts who were sceptical about the durability of the CCP based their views on the expectation that the regime could be undermined by the disappointed hopes and rising expectations of less fortunate members of China's society who had not benefitted greatly from the general prosperity. This indicated a further critical uncertainty about whether the stability of China's authoritarian regime would be threatened from below by social unrest. Several scholars assessed that such unrest was both widespread and growing. Minxin Pei, for example, observed that social tensions – as measured by official statistics for mass incidents and petitions – had risen to record levels. The government's response mechanisms had failed to redress grievances, and its investment-driven economic strategy was causing "crushing social strains" that were "undermining the political stability critical to future prosperity".[27]

Other analysts supported Pei's view. Albert Keidel believed China's modernisation and official corruption had created an "extraordinary scale of social unrest".[28] Jae Ho Chung, Hongyi Lai, and Ming Xia found that social instability in China was "rising unabated" to a level that challenged the authority and legitimacy of the regime.[29] Carl Minzner considered that China's political and social controls were fuelling "a destabilizing cycle of social unrest".[30] Researchers within China also expressed concerns. Yu Jianrong at CASS noted that social conflict was a highly sensitive topic in China and a "veritable mine field" for scholars. He ventured, however, that in the next decade or two China would probably enter "a period of frequent social conflict" in which peasants, workers, and intellectuals would confront the dominant elitist alliance and create political, economic, and social upheaval.[31]

By no means all experts, however, drew such negative conclusions from their study of social protests in China. Kevin O'Brien recognised the scale of such activity, but did not believe rural China was "about to explode". Most protests were weakly organised, circumscribed, small-scale, and local. They had a positive aspect in that they helped to deflect predatory behaviour. There was little evidence they posed "an imminent threat" to the regime. He saw the CCP's tolerance of rural contention as a sign of the Party's confidence and part of its "high-wire legitimation strategy".[32]

Several experts also emphasised that assessments claiming a high level of social unrest in China were contradicted by the results of numerous opinion polls that apparently confirmed high and even strengthening support for the regime. Wenfang Tang noted that a comparison of eleven public opinion polls in Chinese cities between 1987 and 1999 showed that, despite dissatisfaction with the costs of economic development, support for the CCP was high.[33] Zhengxu Wang's analysis of survey data led him to argue that development had generated "substantial public satisfaction" with the regime's performance and a high level of trust. He also claimed that respondents did not appear to be intimidated about expressing their views freely in such polls. They indicated a much higher level of satisfaction with their country's economic development than did respondents in Japan, South Korea, Hong Kong and Taiwan. Wang concluded that in the absence of major crises the regime would remain in power.[34]

Some US analysts were reluctant to accept such survey results. Edward Friedman argued plausibly that respondents in authoritarian societies probably felt compelled to give politically correct answers.[35] However, many others who examined these surveys closely were inclined to differ. Commenting on survey results in the 2003 edition of the CASS *Blue Book of Chinese Society*, Joseph Fewsmith noted that Chinese polls indicated a sizeable majority of citizens were cautiously optimistic about the future. As much as 74 percent rated their overall situation as OK and 73 percent expected that life would be better in the future.[36] Commenting again on the 2007 edition, Fewsmith noted continuing or even deteriorating social problems, but overall optimism persisted. Over 90 percent expressed confidence in the party-state's handling of affairs and thought overall socioeconomic development was good. Overall he thought the data looked "favourable in terms of social stability", although with considerable conflict below the surface and dissatisfaction about specific local issues.[37]

While it might readily be argued that such officially supported surveys could be biased or distorted, other experts conducted their own surveys in China. Based on a survey in 2003, Tony Saich confirmed an apparently high level of satisfaction (86.1 percent) with the central government's performance, although satisfaction with government at the township and village level had declined to 43.6 percent.[38] He repeated these surveys in 2005, 2007, and 2009 and found that levels of satisfaction rose steadily at all levels reaching 95.9 percent for the central government and 61.5 percent for local government. He believed the poorest in Chinese society were responding positively to the Hu-Wen government's social welfare initiatives. Dissatisfaction with local officials and with corruption might be a source of social instability, but citizens did not blame the central government directly.[39] US sociologist Ethan Michelson conducted similar surveys in China's rural counties in 2002 and 2010. He too found growing levels of confidence and trust in government, crediting this to greater provision of social benefits, subsidies, reduced taxes and fees, and infrastructure improvements.[40]

Other positive polling results emerged from the Pew Global Attitudes Survey, which showed that between 2002 and 2008 the percentage of respondents in

China who expressed satisfaction with their country's direction increased from 48 to 86, while the percentage that thought China's economy was good rose from 52 to 82. In the 2011 survey, 85 percent remained satisfied with China's direction and 88 percent saw the economic situation as good. These results were the highest among the 24 countries surveyed, and by a very wide margin.[41]

Such strong findings from opinion polling led two other experts to draw conclusions that differed markedly from the prevailing wisdom about the level of social tension in China. Teresa Wright perceived growing acceptance of authoritarianism in China. She noted that "countless" national polls showed "remarkable support" for the CCP. Its reforms had persuaded most Chinese to accept the status quo. She thought this could change as memories of the Maoist era of deprivation faded and especially if the economy slowed, but even so she believed China might "confound prevalent assumptions that economic liberalization and growth lead to strain between an authoritarian regime and the society that it governs".[42]

Another important critique of prevailing views about social unrest came from Martin Whyte who disputed "the myth of the social volcano" in China. Based on a national survey he conducted in 2004, his assessment was that respondents were critical of aspects of current inequalities, but only in a context of wider acceptance and approval. They perceived that these inequalities – arising when some members of society achieved wealth through hard work and seizing opportunities – were more acceptable than the structural inequalities of the Maoist period that were caused by the rigid assignment of individuals to hierarchies and locations.[43] Whyte also found that China's most disadvantaged citizens (including farmers) were not the ones who were most angry about new patterns of inequality.[44]

Democracy

Despite the seeming durability of China's authoritarianism, a great deal of expert attention has been devoted to assessing the potential for it to adopt a Western-style democracy – much more attention perhaps than would seem justified by the all-too-apparent current obstacles. Several analysts were surprisingly optimistic about the outlook for democratisation. They were also surprisingly specific about the timeframe, variously predicting democratisation between about 2015 and 2030. Early hopes were dealt a major blow by the Tiananmen crackdown, but not completely extinguished. As noted earlier in this chapter, Andrew Nathan was one of those after Tiananmen who still perceived strong forces in favour of democracy, although later developments diluted that view. At the time, however, he observed that "one of the firmest findings" in the social sciences was that a rising middle class would eventually foster an irresistible demand for democratisation. He therefore doubted China's political stability would last, speculating that a future leadership split was likely and questioning whether the "Humpty-Dumpty" of regime legitimacy could be put together again. He thought provincial and even national elections might be introduced, but conceded that fear of instability might induce China to follow a more gradual path to democracy.[45]

Willy Lam was also hopeful. In 1995 he speculated that some Chinese leaders might be attracted to East European-style socialism with "real elections and other elements of pluralistic politics". There would be strong opposition to such changes, but this might be offset by realisation that: "Not budging … might provoke a revolution".[46] Minxin Pei also saw possible progress toward democracy at this time. He thought authoritarian political institutions were fast decaying and hoped that new reforms promoting village-level elections and an expanded role for the National People's Congress (NPC) might lead to "creeping democratization".[47]

Perhaps the most conspicuous of those in the 1990s who saw bright prospects for democracy in China was Henry Rowen, who answered his question: "When will China become a democracy?" with the confident forecast that it would be "around the year 2015". He accepted that some might think such a prediction foolhardy but pointed to "unmistakable signs" of positive change – including the beginnings of grassroots democracy, rule of law, and liberalising of mass media. He asserted that Seymour Lipset had established a "worldwide norm" that the richer a country became the freer it would be. Rowen calculated that by 2015 China's per capita GDP would reach a level at which several scholars had suggested that transition to democracy would become irreversible.[48]

The first of several editions of the *Journal of Democracy* that explored the outlook for China's democratisation was published in 1998. Responding to the question of whether China would still be governed by the CCP in ten years, Zbigniew Brzezinski thought it would be forced by that time to share power. Émigré reformist Yizi Chen thought economic modernisation would undermine CCP rule and lead to electoral democracy within the decade. Arthur Waldron also believed that the CCP would not survive the decade, but thought a democratic outcome unlikely. Most of the other contributors also believed democratic change was only likely in the longer term.[49] Among contributors to another special edition of the journal in 2003, only Bruce Gilley seemed hopeful. He believed the weak institutionalisation of China's authoritarian system would lead to a breakdown that could precipitate democratisation. He supported his argument by drawing attention to another aspect of predictive theory, namely Robert Dahl's invocation of the "logic of concentrated power", to infer the existence of an inexorable drive toward wider political participation.[50]

In following years, Waldron became more optimistic about China's democratic prospects. He reasoned that its dynamic economy and its more affluent and well-informed society could lead to political change sooner than many expected.[51] Gilley also now asserted that conditions in China were shifting decisively in favour of democratic change. He cited de Tocqueville's view that autocratic regimes that left reform until too late only ensured their demise. The CCP, he speculated, could "muddle along for years or disappear in a flash". In a vivid scenario, he suggested 2012 might be the year when China made "a second attempt to build an enduring democracy". Pointing to the theoretical findings of several scholars that democracy would survive "come hell or high water" when per capita incomes rose above a certain level, he calculated that a democratic breakthrough in China would become "impregnable" by around 2020.[52]

Interestingly, a review of Gilley's book that suggested Sinologists should "eschew prognostication and focus on understanding China's complicated present" elicited a lengthy and impassioned response from him, damning this argument as "logically flawed and professionally irresponsible". Predictions, he asserted, were "the mission of social scientists" and without them there would be "a muddle of thought and a muddle of policy". He stood by his forecast that China would obtain democracy by 2020, claiming that even if it were wrong it was useful. Since the idea of democracy was becoming "a universal value", he believed the burden of proof should be on those who argued China would *not* become a democracy.[53]

In 2007, Henry Rowen revised his prediction that China would be free by 2015, claiming it would be "partly free" by 2015 and "free" by 2025. He cited democracy theories developed by Robert Barro and Adam Przeworski who found that higher incomes and education levels predicted higher freedoms. He was, however, prepared to consider that China might arrive at "a novel political arrangement that falls short of true liberal democracy".[54] Dali Yang found it hard to disagree with Rowen's view that China would soon become "partly free". Though some critics dismissed a linkage between per capita income and democracy as "bogus axioms of political development", Yang said regime studies showed dictatorships eventually died as economic development progressed.[55]

Another strong proponent of China's likely democratisation was Cheng Li. He perceived the emergence from the mid-2000s of a populist faction and an elitist faction within the CCP leadership. He believed this "inner-party bipartisanship" would make elections at the higher political level more likely "in the not-too-distant future", and that competition within the CCP "might one day be extended to general elections in the country".[56] On another occasion he assessed that the most likely political scenario for China in 2020 would see "the emergence of a constitutional democracy with Chinese characteristics".[57]

Li accused the China studies community in the US of "political blindness" by overlooking the significance of China's political experiments. He acknowledged that the CCP was not willing to give up its political monopoly or move to a Western-style tripartite division of powers, but believed it realised its legitimacy was being questioned and that it was "moving toward change". The CCP as currently constituted could not survive indefinitely "because societal forces will become increasingly active".[58] Li claimed, in 2009, that leaders who supported competitive elections were now in control in China.[59] His assessment was that, by 2025, inner-party democracy and factional politics would make China's elections more genuine.[60] Yet another analyst who seemed prepared to put an approximate date on China's democratisation was Kerry Brown, who indicated in 2010 that although he was extremely pessimistic about China in the short term, and relatively pessimistic in the medium term, he was "wildly optimistic" in the long term – to the extent that he predicted: "by about 2030 we will be looking at a robust democracy in China".[61]

Such optimism about China's democratic future appeared to be shared, however, by only a minority of China watchers. Others specifically cautioned against overestimating support in China for Western-style democracy. Zheng Yongnian, for example, felt a major issue for China experts was that their emotions "often prevent them from reaching realistic conclusions". He questioned predictive theories that China's embrace of capitalist economic development would lead to democracy, as reflected in the works of scholars such as Charles Lindblom, Samuel Huntington, Seymour Lipset, and Barrington Moore. He shared Andrew Nathan's view that democracy in China was more likely to be gifted by the political elite than created by social forces.[62]

Many others were equally sceptical about China's democratic prospects, including Harry Harding and Thomas Metzger who noted the CCP's firm opposition to full democratisation, the possibility that economic and other setbacks could derail political reform, the vested interests of both rulers and the people in the status quo, and the influence of China's Confucian tradition.[63] Daniel Burstein and Arne de Keijzer did not rule out that in the "far future" China might be attracted by "the Western approach to democracy and personal liberty", but they still considered adoption of full democracy unlikely.[64] Harvey Nelson disputed neo-liberal institutionalist theory that suggested globalisation would favour democratisation.[65] Nathan was also doubtful. His minimum definition for democracy in China included open, competitive, and periodic elections for top power-holders and he did not see this resulting from China's experiments with political reform.[66]

Jonathan Anderson questioned the idea that China could not grow without democracy. He countered that a better question might be whether China could grow *with* democracy, since the other successful Asian economies were also effectively one-party states. He believed there was, in effect, "an unwritten contract" in which the Chinese people agreed to hold off on democratic aspirations in return for the benefits of growth.[67] Even Cheng Li did not always appear hopeful about China's democratic prospects, commenting on one occasion that it should not be expected China would adopt a multi-party system in the near future and that a Chinese democracy would have "its own unique features".[68]

Liberalisation

Apart from uncertainty about whether authoritarian rule would continue to flourish in China or whether Western-style democracy would triumph in its place, another much-debated critical uncertainty – situated somewhere midway between these two alternatives – also aroused the interest of China experts: namely, whether or to what extent political liberalisation might make inroads over time into the nature of the CCP's rule. Within this broader question, two issues in particular attracted much attention in the literature. The first was the extent to which political liberalisation had already progressed within the authoritarian structure and how far it might develop in the future. The second and related issue was whether incremental liberalisation might lead in the long term to the eventual emergence of democracy, and whether this would be Chinese-style or Western-style democracy.

Some analysts seemed impressed by the amount of progress made toward political liberalisation. Even Minxin Pei, at least in 2000, saw signs of "slow but real" political change. As a consequence he was one of the first to use the expression "soft authoritarianism" to describe China's emerging political system, and saw "real potential for democratic transition, perhaps in the second decade of the twenty-first century". He perceived a decline in the use of political repression, increasing political pluralism, strengthening of institutions for rule of law, popular participation, and the development of civil society. He noted the introduction of village elections and an expanded role for the NPC and provincial people's congresses.[69] Others, such as John Thornton, were also encouraged. He said Premier Wen had indicated to a US delegation that he could foresee elections expanding to provincial level.[70] George Gilboy and Benjamin Read perceived the emergence of a "more restrained, rational and rule-governed" political system, as well as more transparency and responsiveness.[71] Steve Tsang also saw indications that the CCP was emphasising governance reform and greater public consultation.[72]

The view taken by many other experts, however, including later by Minxin Pei, was that political liberalisation in China was unlikely to develop much farther because it would threaten the CCP's hold on power. Roderick MacFarquhar, for example, argued that substantial political reform was unlikely because it might remove "the stone that leads to an avalanche".[73] Other experts considered that the degree of political liberalisation the CCP had encouraged or tolerated to date was limited, or that it had stalled or even been reversed. Joseph Fewsmith, for example, commented that the prospects for significant political reform in China were "increasingly dismal". The CCP, in his view, would continue on a trajectory of "modest, inner-party democracy" that would not threaten its control, at least "in the immediate future".[74] Zhengxu Wang pointed to opinion polls that showed that more than 90 percent of Chinese believed democracy was good, but they did not support rapid democratisation because growth and stability were seen as more important.[75] Another poll found that 79 percent were satisfied with "the way democracy works in our country", indicating they believed some form of democracy was already in place.[76]

A mixed assessment of China's political liberalisation was also presented by Charles Freeman and Melissa Murphy. They agreed that the CCP had recognised the need for political reform but thought the trajectory was unclear. The Party favoured greater public participation in political decision-making, but under authoritarian leadership. Its approach was "instrumental, incremental, and idiosyncratic". Its experiments were encouraging, but implementation was problematic.[77] Kerry Brown arrived at similar conclusions in his study of China's village elections, which he described as mired in corruption, vote-rigging, and cronyism.[78]

Clearly, therefore, expert opinion has been divided about the progress of political liberalisation in China. Sanguine views had perhaps been encouraged by seemingly explicit support from some Chinese leaders. Wen Jiabao appeared on several occasions to advocate increasing the momentum of political reform.

In his report to the NPC in 2010 he said China's modernisation and economic reform could risk failure without political restructuring, including expansion of "primary-level democracy". He said the regime intended to "create conditions for the people to criticise and supervise the government and let news media fully play their oversight role".[79] In a speech to the Royal Society in London in 2011 he declared China would be a country that "fully achieves democracy, the rule of law, fairness, and justice". He saw people's democracy as "the soul of socialism" and added "without freedom, there is no real democracy". He also supported "a freer academic atmosphere" characterised by "the spirit of independence and freedom in thinking".[80]

Wen's position certainly appeared to be at the leading edge of liberal political thinking in the CCP but, as Joseph Fewsmith pointed out, there was little evidence that he was advocating a level of reform that would challenge the CCP's power monopoly. His remarks were broadly consistent with the most definitive statement of the CCP's approach to democratisation as reflected in its 2005 White Paper titled *Building of Political Democracy in China.* This document described democracy as a universal aspiration but was adamant that "the political system a country adopts and the road to democracy it takes must be in conformity with the conditions of that country". It endorsed political consultation and participation – including China's system of people's congresses, grassroots democracy and village elections, and inner-Party democracy – as part of "political democracy with Chinese characteristics". But it maintained there was "no one single and absolute democratic mode in the world that is universally applicable". In particular, the White Paper stressed repeatedly that the "most important and fundamental principle" for China's "socialist political democracy" was leadership by the CCP.[81] NPC chairman Wu Bangguo declared that China would never adopt "a multiparty revolving-door system or other Western-style political models". He insisted that "socialism with Chinese characteristics" was "the only correct road" for China.[82]

Some commentators nevertheless saw the prospects for Western-style democracy as still somewhat hopeful, perhaps influenced by the views of prominent Chinese thinkers like Yu Keping, a CCP official and professor at Peking University. Yu's writings – especially his article published in Chinese in 2006 titled "*Minzhu shige haodongxi*" [Democracy is a Good Thing] – attracted much attention.[83] He described democracy as "the best political system" and "the one with the fewest number of flaws". As Andrew Nathan clarified, however, Yu's concept of democracy was much more consistent with the CCP's view than with a minimal Western definition.[84] In his article, Yu advocated "a socialist democracy with unique Chinese characteristics". He did not want to "import wholesale an overseas political model". Elsewhere he wrote tantalisingly of his hope that an incremental approach to democracy in China would lead to "breakthroughs in the fundamental nature of reform", but he suggested the process would be "gradual and long". He preferred an "orderly democracy" whose essence would be "political participation by the citizenry" but within a framework "led by the party and the government". He envisaged that China's citizens would "gradually experience more freedom in political elections" and that these elections would be "more and more competitive", but he saw the CCP remaining "at the center of political power".[85]

The subject of China's democratic prospects was also raised by two Chinese scholars: Liu Yu at Tsinghua University and Chen Dingding at the University of Macau. Despite the "pessimistic view of China's democratic future" held by Western academics, they saw many signs that "a fundamental political transformation" was occurring. In fact they were optimistic that China would become a democracy "around 2020 or so". Importantly, however, they also noted that the actual form that Chinese democracy would take was "uncertain" and was unlikely to resemble the Western model.[86]

Some have nonetheless clung to the hope that at least in the long term China will become fully democratic. The signatories of *Charter 08*, including over 350 intellectuals and human rights activists in China, called for the replacement of authoritarianism by a new political system that would include periodic, competitive elections for all major government positions. However, given the contrary preferences of even China's most liberal leaders and of most of its influential public intellectuals, several experts reached the conclusion that a more-limited Chinese style of democracy was perhaps more likely. Shaohua Hu, for example, believed China would sooner or later be democratic but that it could follow a different model to the West.[87] Even Bruce Gilley, one of those seemingly most confident about China's democratic prospects, speculated that a Chinese-style democracy could be quite different to, and possibly even better than, the Western model. He thought China's democracy could even push established democracies to deepen their own democratic experiences.[88] Henry Rowen also conceded that Chinese democracy could be "a novel political arrangement that falls short of true liberal democracy".[89]

It seems likely that the nature and extent of political liberalisation within the structure of authoritarian rule will continue to be debated as one of the most important critical uncertainties in China's political future. What emerges from the debate so far, however, is that most experts appear to accept that the best that can be expected for a considerable time is a Chinese style of democracy under CCP leadership that will fall well short of what many in the West would define as democracy. This has still, however, not persuaded some China watchers to abandon completely their hopes that true democracy might eventually prevail in the distant future. Cheng Li in 2012 again questioned the durability of the CCP's "authoritarian resilience", arguing that resistance to democratic change is causing the political system to stagnate. To avoid a "bottom-up revolution" it must, he claimed, "embrace a systematic democratic transition with bold steps towards intra-Party elections, judicial independence and a gradual opening of the mainstream media".[90]

Andrew Nathan, in a new 2013 collection of articles in the *Journal of Democracy*, stated that the "consensus" was stronger than at any time since Tiananmen that the CCP's resilience was "approaching its limits". Informed observers perceived deep changes that would eventually have major consequences. Based on his investigation of regime change theories – including threshold models, revolutionary bandwagons and information cascades – Nathan noted that such transitions were paradoxical in that they appeared retrospectively to be "inevitable but not predictable", until a tipping point was reached. Regime change depended not only on there being a majority in favour of change but also on triggering events. In the case of China, a "breakthrough moment" might be triggered by something like "a foreign humiliation". He believed that "the expectation of dramatic change" persisted, and that it could be led by pro-democracy forces inside the CCP. He also conceded, however, that such change might instead "galvanize a shift from social dissatisfaction to social support for the current regime", or it might lead to "a descent into disorder".[91]

Political futures

The most obvious finding from the above consideration of critical uncertainties in China's political domain is that experts expressed a wide range of conflicting views. There was no clear consensus about China's political direction. This concluding section can only summarise some of the various positions in the expert debate. On the question of the prospects for the continuation of an authoritarian regime, a number of experts were quite sceptical but even more were impressed by the strength of the CCP's grip on power and tended to believe it would maintain its legitimacy by responding gradually and as it deemed necessary to popular demands for political participation. They proposed various labels to describe this political model including soft authoritarianism, resilient authoritarianism, flexible authoritarianism, populist authoritarianism and adaptive authoritarianism.[92] As time passed some scholars began again to question these conclusions and to consider the possible impermanence of the authoritarian model, but others countered that in their view the CCP appeared to be increasingly entrenched.

On the question of the prospects for democracy, it is fascinating that, despite the majority view accepting the durability of China's authoritarianism, a significant number of experts appeared very confident that China would in due course achieve substantive democracy. They were even prepared to attach surprisingly specific timeframes to such predictions. However, the majority of other analysts continued to discount the possibility of an imminent Western-style democratic future and to question the predictive power of the various theories, including democratic transition theory, which claimed to support the likelihood of this outcome. The powers of such predictive theory remain at this point distinctly unproven in the Chinese context.

Among the other methods used by experts to forecast China's political future, quantitative analysis also played a part. Those who favoured regime change scenarios frequently cited official Chinese statistics about the rising number of "mass incidents", which they believed indicated a very high level of social unrest. However, several others questioned the significance of these statistics and drew attention instead to the remarkably high level of popular support for the government, as indicated by the findings of numerous opinion polls and surveys.

They refuted "the myth of the social volcano" and identified a high level of acceptance of China's authoritarianism.

Contradictory findings probably contributed to a tendency for other experts to focus on a possible alternative outcome to the triumph of either Western-style democracy or a robust authoritarianism: namely, the possibility of increasing liberalisation within the authoritarian system. Optimists saw promising signs that this was occurring and were encouraged by the apparent support of some Chinese leaders and scholars for accelerated progress toward democracy. But others countered that what most of these Chinese commentators had in mind was a form of "democracy with Chinese characteristics" in which the CCP remained firmly in control. In this situation they maintained that the best that could be hoped for was incremental political liberalisation within an authoritarian structure.

It is evident that many if not most of these diverse expert forecasts about China's political future have failed, at least so far, to be realised. Those who predicted so confidently that democracy would emerge in the short to medium term have been proved wrong, as have those who hoped for greater progress toward political liberalisation. Those who argued for the durability of authoritarianism have been vindicated to date, although the adaptability of this system has not been as great as some may have anticipated.

The expert debate confirms that opinions about the future configuration of China's current authoritarian system, and even more so the prospects for an imagined alternative democratic system, are divided. Discussion will continue about whether the CCP is capable of providing the quality of political governance that its increasingly articulate citizens expect and that their rising expectations of modernity will perhaps demand. It seems clear that the quality and form of political governance in China will remain a critical uncertainty that will continue to be debated hotly among experts both within and outside China, even though most see radical change as unlikely in the short term. The outcome is an open question that will depend on the evolution of the political process and on the influence of key actors both within the leadership and within society. The prevailing uncertainty is well expressed in Hugh White's comment that the real question is whether the CCP can "strike a balance" between keeping control and allowing enough reform for it to retain legitimacy.[93] Political reform seems likely to be incremental at best but there may be scope for the regime to shore up its political legitimacy through progress in areas such as economic and environmental reform, the prospects for which will be considered in the next two chapters.

Notes

1 Minxin Pei, "Is China's Communist Party Doomed?", *The Diplomat* (1 October 2012).
2 Hutton, *The Writing on the Wall*, x.
3 Zheng Yongnian, "Interest Representation and the Transformation of the Chinese Communist Party", in *Bringing the Party Back In: How China is Governed*, ed. Kjeld

Erik Brodsgaard and Yongnian Zheng. (Singapore: Eastern Universities Press, 2004), 297.

4 Zheng Yongnian, *Is Communist Party Rule Sustainable in China?* (China Policy Institute, University of Nottingham, July 2007).

5 Harvey Nelson, "The Future of the Chinese State", in *The Modern Chinese State* David Shambaugh, ed. (Cambridge, UK: Cambridge University Press, 2000), 230–235. Christopher Marsh also argued that China had learned from the Soviet Union: Christopher Marsh, "Learning from your comrade's mistakes: the impact of the Soviet past on China's future", *Communist and Post-Communist Studies* 36 (September 2003).

6 Andrew J. Nathan, "Authoritarian Resilience", *Journal of Democracy* 14, no. 1 (January 2003), 6–17.

7 Ibid., 16. See also: *Fifteen Years After Tiananmen: Is Democracy in China's Future*, (United States Congressional-Executive Commission on China, 3 June 2004).

8 Andrew J. Nathan, *Is Communist Party Rule Sustainable in China?* (Carnegie Endowment for International Peace, 5 October 2006).

9 Kjeld Erik Brodsgaard and Yongnian Zheng, eds. *Bringing the Party Back In: How China is Governed* (Singapore: Eastern Universities Press, 2004), 16–19. See also: Kjeld Erik Brodsgaard and Yongnian Zheng, eds. *The Chinese Communist Party in Reform* (London: Routledge, 2006).

10 Jean-Pierre Cabestan, "Is China Moving Towards 'Enlighted' But Plutocratic Authoritarianism?", *China Perspectives* 55 (September/October 2004), 2–8.

11 Bruce J. Dickson, "Threats to Party Supremacy", *Journal of Democracy* 14, no. 1 (January 2003), 27–35.

12 Bruce J. Dickson, *Populist Authoritarianism: The Future of the Chinese Communist Party* (Carnegie Endowment for International Peace, 2 November 2005).

13 Bruce J. Dickson, *Red Capitalists in China: The Party, Private Entrepreneurs, and Prospects for Political Change* (Cambridge, UK: Cambridge University Press, 2003), 2–16 and 157–171.

14 Bruce J. Dickson, *Wealth into Power: The Communist Party's Embrace of China's Private Sector* (Cambridge, UK: Cambridge University Press, 2008), and Chen Jie and Bruce J. Dickson, *Allies of the State: China's Private Entrepreneurs and Democratic Change* (Boston: Harvard University, 2010). See also Kellee S. Tsai, *Capitalism Without Democracy: The Private Sector in Contemporary China* (Ithaca, NY: Cornell University, 2007).

15 James Mann, *The China Fantasy: How our Leaders Explain Away Chinese Repression* (New York: Viking, 2007).

16 Shambaugh, *China's Communist Party*, 1–9 and 168–178.

17 David Shambaugh, "China's Communist Party at 90", *International Herald Tribune* (30 June 2011).

18 Andrew J. Nathan, "Authoritarian Impermanence", *Journal of Democracy* 20, no. 3 (July 2009), 37–40.

19 Richard McGregor, *The Party: The Secret World of China's Communist Leaders* (New York: HarperCollins, 2010), 273.

20 Richard McGregor, "5 Myths About the Chinese Communist Party: Market-Leninism Lives", *Foreign Policy* (January/February 2011).

21 Zheng Yongnian, *The Chinese Communist Party as Organizational Emperor: Culture, Reproduction and Transformation* (London: Routledge, 2010).

22 Kerry Brown, *Friends and Enemies: The Past, Present and Future of the Communist Party of China* (London: Anthem, 2009), 23.

23 John Wong and Hongyi Lai, eds. *China into the Hu-Wen Era: Policy Initiatives and Challenges* (Singapore: World Scientific, 2006), 25–29.

24 Lam, *Chinese Politics in the Hu Jintao Era*, ix–xiii.

25 Eric C. Anderson, *China Restored: The Middle Kingdom Looks to 2020 and Beyond* (Santa Barbara, CA: Praeger, 2010), 117 and 243–246.

26 Stephen Platt and Jeffrey Wasserstrom "China's Long History of Defying the Doomsayers", *The Atlantic* (30 August 2012). See also Eric X. Li, "The Life of the Party", *Foreign Affairs* 92, no. 1 (January/February 2013), 34–46, and Yasheng Huang, "Democratize or Die", *Foreign Affairs* 92, no. 1 (January/February 2013), 47–54.
27 Minxin Pei, "China is Paying the Price of Rising Social Unrest", *Financial Times* (7 November 2005).
28 Albert Keidel, *The Economic Basis for Social Unrest in China* (Third European-American Dialogue on China, George Washington University, 26–27 May 2005). See also Albert Keidel, *China's Social Unrest: The Story Behind the Stories* (Carnegie Endowment for International Peace, September 2006).
29 Jae Ho Chung, Hongyi Lai, and Ming Xia, "Mounting Challenges to Governance in China: Surveying Collective Protestors, Religious Sects and Criminal Organizations", *The China Journal* 56 (July 2006), 1–2 and 30.
30 Carl Minzner, "Social Instability in China: Causes, Consequences and Implications", in *The China Balance Sheet in 2007 and Beyond*, ed. C. Fred Bergsten, Bates Gill, Nicholas R. Lardy and Derek J. Mitchell. (Washington, DC: Center for Strategic and International Studies and the Peterson Institute for International Economics, May 2007), 55–77.
31 Yu Jianrong, "Social Conflict in Rural China", *China Security* 3, no. 2 (Spring 2007), 2–17.
32 Kevin J. O'Brien, "Rural Protest", *Journal of Democracy* 20, no. 3 (July 2009), 25–28.
33 Wenfang Tang, *Public Opinion and Political Change in China* (Stanford, CA: Stanford University Press, 2005), Chapters 3 and 8.
34 Zhengxu Wang, "Explaining Regime Strength in China", *China: an International Journal* 4, no. 2 (September 2006), 217–237, and Zhengxu Wang, "Citizens' Satisfaction with Government Performance in Six Asia–Pacific Giants", *Japanese Journal of Political Science* 11, no. 1 (2010).
35 Gang Lin, ed. *China's "Credibility Gap": Public Opinion and Instability in China* (Washington, DC: Woodrow Wilson International Center for Scholars, August 2002), 17–18.
36 Joseph Fewsmith, "China's Domestic Agenda: Social Pressures and Public Opinion", *China Leadership Monitor* 6 (Spring 2003).
37 Joseph Fewsmith, "Assessing Social Stability on the Eve of the 17th Party Congress", *China Leadership Monitor* 20 (Winter 2007). See also Joseph Fewsmith, "The Political Implications of China's Growing Middle Class", *China Leadership Monitor* 21 (Summer 2007).
38 Tony Saich, "Citizens' Perceptions of Governance in Rural and Urban China", *Journal of Chinese Political Science* 12, no. 1 (2007), 1–28.
39 Tony Saich, "Governance Seen through the People's Eyes,, *East Asia Forum Quarterly* (April–June 2011), 7–8.
40 John Sexton, *China Rural Unrest on the Wane, Survey Claims* (china.org.cn, 22 March 2011).
41 *The 2008 Pew Global Attitudes Survey in China: The Chinese Celebrate their Roaring Economy, As they Struggle with its Costs* (Pew Global Attitudes Project, 22 July 2008), and *China Seen Overtaking US as Global Superpower: 23-Nation Pew Global Attitudes Survey*, (Pew Research Center Global Attitudes Project, 13 July 2011).
42 Teresa Wright, *Accepting Authoritarianism: State–Society Relations in China's Reform Era* (Palo Alto, CA: Stanford University Press, 2010), 162–179.
43 Martin King Whyte, "Myth of the Social Volcano: Popular Responses to Rising Inequality in China", in *The People's Republic of China at 60*, ed. William C. Kirby (Cambridge, MA: Harvard University Press, 2011). See also Martin King Whyte, *Myth of the Social Volcano: Perceptions of Inequality and Distributive Injustice in Contemporary China* (Palo Alto, CA: Stanford University Press, 2010).

44 Martin King Whyte and Maocan Guo, "How Angry are Chinese Citizens about Current Inequalities? Evidence from a National Survey", in *Social Stratification in Chinese Societies*, ed. Kwok-bun Chan, Agnes S. Ku, and Yin-wah Chu (Leiden: Brill, 2009).
45 Andrew J. Nathan, "China's Path from Communism", *Journal of Democracy* 4, no. 2 (April 1993), 30–42.
46 Lam, *China after Deng Xiaoping*, 383–430.
47 Minxin Pei, "'Creeping Democratization' in China", *Journal of Democracy* 6, no. 4 (October 1995), 65–79.
48 Henry S. Rowen, "The Short March: China's Road to Democracy", *The National Interest* 45 (Fall 1996).
49 "Will China Democratize?", *Journal of Democracy* 9, no. 1 (January 1998).
50 Bruce Gilley, "The Limits of Authoritarian Resilience", *Journal of Democracy* 14, no. 1 (January 2003).
51 Arthur Waldron, "How Would Democracy Change China?", *Orbis* 48, no. 2 (Spring 2004).
52 Bruce Gilley, *China's Democratic Future: How It Will Happen and Where It Will Lead* (New York: Columbia University Press, 2005), 28–44, 60, 97–98 and 151–163.
53 Bruce Gilley, "Should We Try to Predict Transitions to Democracy?: Lessons for China", *Whitehead Journal of Diplomacy and International Relations* (Winter/Spring 2005).
54 Henry S. Rowen, "When Will the Chinese People Be Free?", *Journal of Democracy* 18, no. 3 (July 2007), 38–52.
55 Dali L. Yang, "China's Long March to Freedom", *Journal of Democracy* 18, no. 3 (July 2007), 58–64.
56 Cheng Li, *China's Changing Political Landscape: Prospects for Democracy* (Washington, DC: Brookings Institution, 2008), 14 and 115.
57 Cheng Li, "China in the Year 2020: Three Political Scenarios", *Asia Policy* 4 (July 2007), 18–29.
58 Cheng Li, "China's Political Trajectory: Internal Contradictions and Inner-Party Democracy", in *Global Giant: Is China Changing the Rules of the Game?*, ed. Eva Paus, Penelope B. Prime, and Jon Western (New York: Palgrave Macmillan, 2009), 51–72.
59 Cheng Li, "Intra-Party Democracy in China: Should We Take It Seriously?", *China Leadership Monitor* 30 (Fall 2009).
60 Cheng Li, "Democracy Gaining Momentum in China", *San Francisco Chronicle* (4 June 2009).
61 Brown, "China Will Get Democracy".
62 Zheng Yongnian, *Will China Become Democratic?: Elite, Class and Regime Transition* (Singapore: Eastern Universities Press, 2004).
63 "Will China Democratize?", 12–25.
64 Burstein and de Keijzer, *Big Dragon: the Future of China*, 226 and 316.
65 Harvey Nelson, "Caution: Rough Road Ahead", in *What If China Doesn't Democratize?: Implications for War and Peace*, ed. Edward Friedman and Barrett L. McCormick (Armonk, NY: M. E. Sharpe, 2000), 260–261.
66 Andrew J. Nathan, "China's Political Trajectory: What Are the Chinese Saying?", in *China's Changing Political Landscape: Prospects for Democracy*, ed. Cheng Li (Washington, DC: Brookings Institution, 2008).
67 Minxin Pei and Jonathan Anderson, "The Color of China", 24.
68 Hoffmann and Enright, *China into the Future: Making Sense of the World's Most Dynamic Economy*, 263. Daniel Bell has argued that the idea of the CCP being succeeded by a Chinese Confucian Party was probably less far-fetched than transition to a Western-style democracy. See: Daniel Bell "The Chinese Confucian Party?", *Globe and Mail (Toronto)* (26 March 2010).

69 Minxin Pei, "China's Evolution Toward Soft Authoritarianism", in *What If China Doesn't Democratize: Implications for War and Peace*, ed. Edward Friedman and Barrett L. McCormick (New York: M. E. Sharpe, 2000).
70 John L. Thornton, "Long Time Coming: The Prospects for Democracy in China", *Foreign Affairs* 87, no. 1 (January/February 2008), 2–22. Thornton conceded, however, that China's idea of democracy differed from that of the West and that how far it would develop was an open question.
71 George J. Gilboy and Benjamin L. Read, "Political and Social Reform in China: Alive and Walking", *The Washington Quarterly* 31, no. 3 (2008), 143–164.
72 Steve Tsang, "Consultative Leninism: China's new political framework", *Journal of Contemporary China* 18, no. 62 (November 2009), 865–880.
73 Verna Yu, "Reform unlikely, says China expert Roderick MacFarquhar", *South China Morning Post* (31 October 2012).
74 Joseph Fewsmith, "Political Reform was Never on the Agenda", *China Leadership Monitor* 34 (February 2011). See also Joseph Fewsmith, *The Logic and Limits of Political Reform in China*, (New York: Cambridge University Press), 2013.
75 Zhengxu Wang, "Public Support for Democracy in China", *Journal of Contemporary China* 16, no. 53 (November 2007), 561–579.
76 Nathan, *Is Communist Party Rule Sustainable in China?*
77 Charles Freeman and Melissa Murphy, "Democracy with Chinese Characteristics: Political Reform and the Future of the Chinese Communist Party", in *China's Rise: Challenges and Opportunities*, ed. C. Fred Bergsten, *et al.*, 57–74.
78 Kerry Brown, *Ballot Box China: Grassroots Democracy in the Final Major One-Party State* (London: Zed Books, 2011).
79 *Xinhua* (6 March 2011).
80 *Xinhua* (28 June 2011).
81 China.org.cn, 19 October 2005.
82 Bao Daozu, "Top legislator warns of chaos unless correct path is taken", *China Daily* (11 March 2011).
83 This and other articles by Yu were translated in Yu Keping, *Democracy is a Good Thing: Essays on Politics, Society and Culture in Contemporary China* (Brookings Institution, 2009).
84 Nathan, "China's Political Trajectory".
85 Yu Keping, "Ideological Change and Incremental Democracy in Reform-Era China", in *China's Changing Political Landscape: Prospects for Democracy*, ed. Cheng Li (Washington, DC: Brookings Institution, 2008), 44–58.
86 Liu Yu and Chen Dingding, "Why China Will Democratize", *The Washington Quarterly* (Winter 2012).
87 Hu, *Explaining Chinese Democratization*, 157.
88 Gilley, *China's Democratic Future*, 246–247.
89 Rowen, "When Will the Chinese People Be Free?", 49.
90 Cheng Li, "The End of the CCP's Resilient Authoritarianism? A Tripartite Assessment of Shifting Power in China", *China Quarterly* 211 (September 2012), 595–623.
91 Andrew J. Nathan, "China at the Tipping Point?: Foreseeing the Unforeseeable", *Journal of Democracy* 24, no. 1 (January 2013), 20–25.
92 See also Jessica Teets' contention that "consultative authoritarianism" is emerging in China: Jessica C. Teets, "Let Many Civil Societies Bloom: The Rise of Consultative Authoritarianism in China", *China Quarterly* 213 (January 2013), 19–38.
93 White, "Power Shift: Australia's Future Between Washington and Beijing", 16–17.

Bibliography

Anderson, Eric C. *China Restored: The Middle Kingdom Looks to 2020 and Beyond.* Santa Barbara, CA: Praeger, 2010.

Brown, Kerry. *Friends and Enemies: The Past, Present and Future of the Communist Party of China.* London: Anthem, 2009.

Dickson, Bruce J. *Populist Authoritarianism: The Future of the Chinese Communist Party.* Carnegie Endowment for International Peace, 2 November 2005.

Fewsmith, Joseph. *The Logic and Limits of Political Reform in China.* New York: Cambridge University Press, 2013.

Freeman, Charles and Melissa Murphy. "Democracy with Chinese Characteristics: Political Reform and the Future of the Chinese Communist Party". In *China's Rise: Challenges and Opportunities*, edited by C. Fred Bergsten *et al.*

Gilley, Bruce. *China's Democratic Future: How It Will Happen and Where It Will Lead.* New York: Columbia University Press, 2005.

Li, Cheng. *China's Changing Political Landscape: Prospects for Democracy.* Washington, DC: Brookings Institution, 2008.

Mann, James. *The China Fantasy: How Our Leaders Explain Away Chinese Repression.* New York: Viking, 2007.

McGregor, Richard. *The Party: The Secret World of China's Communist Leaders.* New York: HarperCollins, 2010.

Nathan, Andrew J. "China's Political Trajectory: What Are the Chinese Saying?" In *China's Changing Political Landscape: Prospects for Democracy*, edited by Cheng Li. Washington, DC: Brookings Institution, 2008.

Nathan, Andrew J. *Is Communist Party Rule Sustainable in China?* Carnegie Endowment for International Peace, 5 October 2006.

Shambaugh, David. *China's Communist Party: Atrophy and Adaptation.* Berkeley, CA: University of California, 2008.

Whyte, Martin King. *Myth of the Social Volcano: Perceptions of Inequality and Distributive Injustice in Contemporary China.* Palo Alto, CA: Stanford University Press, 2010.

Wright, Teresa. *Accepting Authoritarianism: State–Society Relations in China's Reform Era.* Palo Alto, CA: Stanford University Press, 2010.

Yu Keping. *Democracy Is a Good Thing: Essays on Politics, Society and Culture in Contemporary China.* Brookings Institution, 2009.

Zheng Yongnian. *The Chinese Communist Party as Organizational Emperor: Culture, Reproduction and Transformation.* London: Routledge, 2010.

Zheng Yongnian. *Is Communist Party Rule Sustainable in China?* China Policy Institute, University of Nottingham, July 2007.

Zheng Yongnian. *Will China Become Democratic?: Elite, Class and Regime Transition.* Singapore: Eastern Universities Press, 2004.

5 Economic uncertainties

The highest-level critical uncertainty for China's future in the economic domain can probably best be expressed simply as: *What are the prospects for China's economy?* Within this over-arching critical uncertainty, several others can be readily identified at the next level down in a hierarchy of uncertainties. Three of the most important of these have received a great deal of attention in the forecasting endeavours of China experts, and are therefore considered in this chapter. Since the beginning of the post-Mao period, one question in particular dominated the debate: namely, *How fast will China's economy grow?* In recent decades this critical uncertainty has frequently taken the additional and more specific form: *When will China's economy overtake the United States?* A third and perhaps more important critical uncertainty, relating to the *quality* of growth, lies behind these two quantitative issues: namely, *Will China be successful in transitioning to a more sustainable economic model?* In recent years this last question has become the main preoccupation among economists.

Economic growth

It is not surprising that China experts have been fascinated by the question of its likely future growth rate. No other large country had expanded as rapidly as China did from the early-1980s. The economic impact of that growth, both domestically and internationally, soon became a subject of great interest. That interest was heightened by the fact that China's economic growth would be a major factor underpinning its success in other domains – including its political legitimacy, its capacity to address environmental concerns, and its ability to exert influence internationally.

Forecasting of China's economic growth was naturally an area of expert analysis in which quantitative techniques received primary emphasis, relying especially on statistical modelling and econometrics. Many organisations within the Chinese government and academia – in collaboration with international partners and with each other – devoted intense effort, over several decades, to assembling and modelling the relevant data. Typically their work was based on variations of the classical production function used to calculate GDP growth. They applied various assumptions or estimations to calculate the rate of growth of the three main factors of production – labour, capital and productivity – which in turn allowed them to estimate the growth of GDP.[1]

This was not an area, however, where econometric forecasting enjoyed notable success. The major distinguishing feature of the majority of forecasts of China's economic growth over several decades was their consistent underestimation.[2] Even those who thought they were being optimistic often underestimated both the strength and durability of growth. These optimists, as already noted in Chapter 2, included analysts in the 1980s like Herman Kahn and Dwight Perkins who expected growth to continue at around 7 percent or higher for the remainder of the twentieth century. They based this expectation on China's apparent commitment to economic reform and the argument that its growth path would therefore probably follow that of other successful East Asian economies like Japan, South Korea and Taiwan. Others were more cautious, although quantitatively their forecasts were often quite similar, with estimates again hovering around the apparently alluring 7 percent mark.

The World Bank offered an excellent example of a major economic forecaster that maintained over time a cautious and conditional but still relatively positive position regarding China's growth prospects. A 1985 in-country study conducted by the Bank concluded that GDP growth approaching 7 percent was feasible, although it also considered a slower growth scenario. It wisely emphasised that it had made projections not predictions and that forecasts were "often confounded", especially in a country like China.[3] A new report after Tiananmen presented a quite positive set of scenarios. The base scenario assumed a "somewhat accelerated schedule" of reform, leading to annual growth of 7.5 percent to 2000. A rapid reform scenario projected annual growth of 8.5 percent. But it warned that calls to increase growth to 10 percent (nearer to the actual outcome) would discourage reform and create a "stop–go cycle".[4] Another Bank forecast in 1997 projected that GDP growth would slow from 8.4 percent in 1996–2000 to 6.9 percent in 2001–2010 and 5.5 percent in 2011–2020, although it did include a comment that sustained rapid growth to 2020 was "clearly possible". With further reforms and skilled management this might lead to "the most remarkable economic transformation the world has ever seen". But it also had a bet the other way. Its main concern was not with defending the Bank's actual record of underestimation but with responding to accusations of over-optimism. It cautioned that there was "always the possibility of nonlinear change" and that China's nonlinearities might be "especially pronounced". If reforms faltered this could lead to "Sinoclerosis".[5]

The World Bank continued in later years to struggle with the task of assessing China's future economic prospects. A major study in 2007 noted "certain fragilities", but concluded that these need not be serious constraints on growth. Nonetheless its projections, which it accepted were conservative, indicated GDP would grow by an average of only 6.6 percent between 2005 and 2020.[6] In 2009, however, its China Office issued a revised and significantly less cautious estimate authored by Louis Kuijs. This projected that, despite the onset of the GFC, growth would average 8.4 percent from 2010 to 2015, and 7.0 percent from 2016 to 2020.[7] Even more optimistically, the Bank's Chief Economist Justin Yifu Lin later commented that China could maintain 8 percent annual growth until 2030 "if the country can overcome various challenges".[8]

Early projections of China's GDP by RAND were much more conservative than even the World Bank. From 1989, a team led by Charles Wolf issued several pessimistic long-term growth forecasts. The first report projected GDP (PPP) growth at only 4.7 percent between 1990 and 2010.[9] Another 1995 report projected annual growth to 2015 at 4.9 percent in a "stable growth" scenario and 3 percent in a "disrupted growth" scenario.[10] Surprisingly RAND failed to learn from its earlier excessive pessimism, issuing two further reports in 2000 and 2001 both of which continued to project stable growth of around 5 percent and disrupted growth of around 3 percent up to 2020.[11]

Despite persistent underestimation by the World Bank and RAND, from the mid-1990s the credibility of China's continuing economic rise was reinforced by increasingly positive growth estimates from several other international organisations and economic forecasters. The Organisation for Economic Cooperation and Development (OECD) in 1996 considered three possibilities for the following 15 to 20 years: an optimistic outcome with continuation of a 9 percent growth trajectory, a less favourable performance causing growth to fall to 4 or 5 percent, and a mixed outcome leading to growth of only 5 to 6 percent. The "prevailing view" was that continued high growth (around 8 percent) was likely. This optimistic view proved of course to be not far below the actual outcome, but OECD analysts were spectacularly mistaken in predicting that China was unlikely to run long-term surpluses in either its trade balance or balance of payments, and in questioning its ability to attract foreign investment.[12] Another OECD report in 2007 concluded that China's high growth rate was "unsustainable" over the long term due to demographic trends, but still not expected to fall much below 7 to 8 percent by 2025.[13]

A 1997 report to the US Congress also included a distinctly optimistic forecast for China's economy to 2010. It warned of likely strains arising from a contradiction between centralised control and a free-wheeling market system, but expected China's GDP to increase by an average of over 9 percent in 1995–2005 and over 8 percent in 2005–2010. This was one of the most optimistic forecasts during this period, and therefore one of the closest to the (even higher) actual outcome. Moreover, the report noted that even such impressive straight-line projections to 2010 might still "seriously understate" future performance potential, but only if accompanied by transition to a democratic market system.[14]

Additional increasingly optimistic projections in the early-2000s by investment advisers Deutsche Bank, Goldman Sachs, and Morgan Stanley have already been noted in Chapter 2, although they also were outstripped by China's actual performance. Not all were convinced, however. Murray Weidenbaum in 1997 suggested that the only forecast that could be made with confidence was that "a straight-line extrapolation of the past is most unlikely to occur". China, he declared, would "not continue to grow at double-digit rates for the coming decade".[15] Thomas Rawski, in 2003, thought poor decisions would keep China on a path of low growth. He saw a lesson for China watchers in Japan's declining growth and in the "rapid obsolescence" of books about Japan's economic ascendancy, illustrating "the twin dangers of overconfidence and linear projections".[16] He later warned that fiscal mismanagement could "tip China's dynamic economy into Japanese-style quicksand".[17]

Despite such dissenters, other experts on China's economy continued to be much more optimistic. John Wong was among the most positive, believing China had become the world's most successful transitional economy and that it was "easy to be optimistic" about its future potential. Its dynamic growth could be "slowed but not stifled".[18] Nicholas Lardy also considered that China's economic prospects were "far better than the pessimists believe". Its rapidly expanding private sector would leave them "scrambling to find something else to worry about instead".[19] Martin Wolf offered the novel opinion that China was underachieving and not living up to its full potential. If it would only adopt the right policies, its growth would accelerate and its economic boom could be "in its middle, not its end". What was remarkable, he felt, was not how quickly China had grown but "how slowly it has done so".[20]

During the second half of the 2000s, there was no shortage of other experts who also took a very sanguine view of China's economic prospects. The opinions of Robert Fogel, Albert Keidel, Ross Garnaut, and Hu Angang have already been cited in Chapter 2. Fogel, who predicted growth of 8 percent or more up to 2040, was encouraged by China's continued reforms.[21] Keidel believed there was little reason to be as pessimistic as some experts because China's leaders were "agile, energetic" and were pursuing a "hive of reform proposals" in a political system that was "nimble and robust".[22] Barry Naughton, one of the most eminent China economists, appeared to agree. As noted in Chapter 3, he saw a "very consistent pattern" that whenever problems in China got "really dramatic" the policy-makers would address them. Many analysts had predicted the collapse of China's growth, the demise of the CCP, and the discrediting of the China model, but so far these sceptics had "never been right".[23]

Another upbeat economist was Dwight Perkins, who complained that optimists were "generally looked down upon in the academic world". It was easier to be a pessimist if you wanted peer respect. Like Naughton his optimism was reinforced by long observation that, when steps were needed, China's leaders had "eventually gone ahead and made them".[24] He assessed that "no plausible pace of economic reform" could reduce growth to below 5 percent.[25] But Perkins' optimism was actual quite conservative. He expected only 6–8 percent growth during the decade to 2015, falling to 5–7 percent in the decade to 2025.[26]

It is noteworthy that growth forecasts by economists inside China tended often, and over a long period, to be more optimistic than the majority of their Western counterparts. In a 1987 book by Chinese economists, CASS Vice-President Liu Guoguang noted that the government's target to quadruple per capita GDP by 2000 required annual GDP growth of 7.2 percent, but he expressed confidence that "a still greater speed" would be achievable. Looking beyond 2000 the authors considered with commendable prescience that China's economy would "continue to forge ahead". They expected that the mid-2000s would see "tremendous change" that would "literally put China in the front ranks of the countries of the world".[27] A decade later, Li Shantong and Fan Zhai in the State Council's Development Research Center (DRC) presented a business-as-usual scenario, forecasting average annual growth of 7.9 percent from 2000 to 2010 and 6.6 percent from 2010 to 2020.[28] Cai Fang and Wang Meiyan in CASS noted doubts among Western economists about the sustainability of China's growth, but considered that rapid growth would continue to be "unassailable" over coming decades.[29]

Guo Wanda, in Shenzhen, observed that there were four different opinions among Chinese economists. The first group thought problems might emerge after 2010, the second that an aging population would slow the economy after 2020, the third that rapid growth would continue until 2030, while the fourth and most optimistic group thought rapid growth would last until about 2050. Guo agreed with the second and third groups, which he said formed the "mainstream" among Chinese economists. He highlighted a number of possible "hiccoughs" or risks, but felt they could be resolved.[30] Similarly Fu Jun at Peking University thought that, "barring disastrous events such as wars", China's growth rate would continue at about 8 percent for the next ten to 20 years.[31] A confident view was also taken by Wang Yahua and Hu Angang, who highlighted that China's growth performance had exceeded "almost all expectations".[32] A collaborative study in 2007 between Chinese and US researchers was equally positive. A business-as-usual scenario projected 8.1 percent growth between 2006 and 2020, an optimistic scenario projected 8.6 percent, and a risk scenario projected 7.1 percent.[33]

Overtaking the United States

As rapid growth continued and China's economy grew progressively larger a new uncertainty began to fascinate China watchers: namely, when – if ever – would China's economy surpass that of the United States. This speculation marked something of a watershed in the development of both expert and popular perceptions about China's potential future status. Prior to the turn of the century, few analysts mentioned this possibility. One of the earliest examples appeared in 1997 in an Australian government-sponsored research paper. The paper drew on work by Angus Maddison, Harry Wu and others that measured GDP in terms of purchasing power parity, which led them to believe that by 1990 China had already become the world's second largest economy. This in turn raised "the question of when China's GDP in PPP terms will overtake that of the USA". Their answer was "sometime between 2010 and 2035". More specifically, on the assumption that China's GDP would grow annually by 7 percent and the US by 2.7 percent, they estimated China's GDP (PPP) would overtake the US in 2020.[34]

Such predictions sparked an enduring controversy about the measurement and validity of comparisons between the economies of China and the US in PPP terms that has still not been resolved satisfactorily. As the Australian paper noted, the advantage of the PPP approach is that it uses converters that "better reflect the ability of a unit of local currency to purchase goods and services in its country of issue". Comparisons based on fluctuating market exchange rates (MER) only accurately reflected the comparative value of goods traded internationally. Other experts have countered that some of the key ingredients of national power, such as advanced military technology and hardware, are best reflected by MER. They pointed out that there are "several problems" with the methodology and practice of producing, and especially updating, PPP converters.[35] Nevertheless, it seems incontestable that the normal practice of making GDP country comparisons in US dollars at current exchange rates is seriously misleading because it ignores that fact that many important components of GDP – for example transportation, construction, infrastructure, and other labour-intensive activities – cost much less in a country like China than in the US.

Another early study by Chinese scholars Hu Angang and Men Honghua noted China's GDP (PPP) in 2000 was half that of the US, and (like the Australian paper) estimated that by 2020 it would be greater than the US. They also sought to compare the two countries in terms of Comprehensive National Power (CNP), taking into account relative national strength in eight key resource categories. They estimated that US CNP in 1998 was three times China's, but by 2020 would be only double.[36] A further early study by Emilio Casetti, using PPP data, contained seven separate projections, all of which indicated that China would achieve economic parity with the US before 2050, with four indicating parity before 2020.[37]

Progressively more optimistic PPP-based estimates of China's prospects continued to appear, all insisting PPPs were more appropriate for comparing the total size of national economies. Morgan Stanley projected in 2004 that China's GDP (PPP) should be bigger than the United States within a decade.[38] *The Economist* estimated that China would overtake the US in 2017.[39] Carsten Holz estimated this would occur between 2012 and 2015.[40] A major new report by Angus Maddison in 2007 also projected that China was likely by 2015 in PPP terms to resume its historic position, during the tenth to the fifteenth century, as the world's largest economy. He estimated that in 2003 China's economy had already reached 74 percent of the US and was more than twice as large as Japan, which it overtook in 1992.[41]

By the late 2000s, however, the use of PPP measures had become increasingly controversial, giving rise to some heated exchanges. In large part this reflected growing doubt about the reliability of PPP figures due to the huge discrepancies between various PPP-based estimates. In 2008, for example, Maddison and Wu noted that the World Bank had recently issued new multinational PPP comparisons that indicated that China's GDP in 2005 was only 43 percent of the US, compared with their own estimate of 82 percent. They found the Bank's "weird results" to be "highly implausible".[42] In response two World Bank economists were equally derisory of Maddison and Wu's views, which they believed were "cavalier" and "naïve". They defended the Bank's "robust multilateral comparison process" and rejected Maddison and Wu's "idiosyncratic and increasingly anachronistic" PPPs.[43]

The World Bank's radical revision of its PPP estimate for China's GDP attracted considerable attention from other experts also. Albert Keidel noted that this 40 percent reduction of China's GDP meant that alarming estimates that it would overtake the US were now "moot". China, he commented, was "just not that big now and will not get big any time soon".[44] *The Economist* confirmed that the difficulty in measuring PPP was one reason why economists preferred MER, but it added that, though PPP data might be imperfect, it still gave "a better picture of the relative size of economies than market exchange rates do".[45] Frederic Vogel, who had overseen the update, agreed that some of its assumptions could be challenged but he saw it as a useful benchmark. A spokesman from the University of Pennsylvania, which maintains a reputable alternative set of international PPP estimates, added that its own figures would also be revised.[46] These various reports drew attention to the high level of disagreement and confusion surrounding the estimation of PPPs, but they also reflected continuing support for an alternative to misleading MER-based comparisons.

Meanwhile projections based on PPPs continued to produce ever nearer horizons for China's economy to overtake the US. Journalists noted that the IMF's updated *World Economic Outlook* database indicated China would overtake the US in 2016.[47] US economist Robert Feenstra estimated China would overtake the US by 2014.[48] The Conference Board estimated this would occur in 2012.[49] Justin Lin reportedly stated that, in PPP terms, China's economy would be twice the size of the US by 2030.[50] One analyst trumped all of these projections, however. Arvind Subramanian at the Peterson Institute asked whether China might not *already* be the world's number one economy in PPP terms. By his calculations, China had overtaken the US some time during 2010. He too considered PPP estimates were a "more revealing way of comparing standards of living and economic size across countries".[51]

Subramanian sought to expose a "central conceit" which was that US economic pre-eminence would not be seriously threatened. In his book *Eclipse: Living in the Shadow of China's Economic Dominance*, he presented an "index of dominance" based on three key criteria: namely, GDP, trade, and the extent to which a country is a net creditor. Assuming 7 percent growth in China and 2.5 percent in the US, he estimated that by 2030 China's GDP (measured half at MER and half at PPP) would be one third greater than the US. Its trade would be twice that of the US. China would be a substantial net creditor and the US a debtor. Subramanian claimed that already in 2010 the US had only a marginal one percent advantage when measured by his index of dominance. He predicted that by 2030 further decline in the US would give rise to a "near-unipolar world" dominated by China. The gap between China and the US in 2030 would be similar to that between the US and its rivals in the 1970s.[52] In his book he also predicted that the Chinese *yuan* could "rival the dollar" as a reserve currency by 2020.[53]

Leaving aside for the moment all these estimates based on PPPs, it is time to shift attention to the fact that over the past two decades a competing set of estimates about China's prospects for overtaking the US economy has appeared in parallel that were all based on market exchange rate comparisons. These MER-based forecasts received more attention and seemed to have more impact that those based on PPP, despite their underestimation of the strength of less developed economies. Initially, expectations of China's performance in MER-based forecasts were quite modest. It was not expected to rival the US until 2050 or later.[54] Many subsequent forecasts during most of the 2000s also remained very pessimistic. Fu Jun thought that, even with an average annual growth rate of 8 percent, China would not overtake the US economy at MER until 2040.[55] Richard Cooper at Harvard University calculated that in 2025 China's GDP would still be only one third of the US.[56] *The Economist* was also initially cautious, estimating that by 2026 China's GDP (MER) would still be less than half that of the US.[57] Similarly, World Bank economists considered that in 2020 China's GDP (MER) would probably be less than 30 percent of the US.[58]

Some concluded that China's economy might *never* surpass the US because the latter simply had "too big a lead".[59] John Tatom in 2007 argued that speculation about China's economic ascendency was "wildly premature", and unlikely to happen "for another thirty years or more".[60] The Economist Intelligence Unit considered it too early to talk of "Asia's century" because the US would remain the most important single country "across all the dimensions of power".[61] Derek Scissors at the Heritage Foundation argued as recently as 2010 that there was "a reasonable chance" that China's economy would never surpass that of the US, given the potential for the US economy to revive and for China's to decline.[62]

Against the weight of such scepticism, however, a new forecast was offered in 2006 by John Hawksworth at PricewaterhouseCoopers. In the first of several papers about the prospects for the major emerging economies to 2050, he pointed to the likelihood that PPPs and MERs would steadily converge over that timeframe. He estimated initially, and very cautiously, that by 2050 China would only reach 95 percent of the US economy at MER and be 40 percent larger in PPP.[63] Two years later, however, he revised his forecasts drastically, concluding that China would overtake the US economy at MER by 2025 and be 29 percent larger in 2050 measured in either PPP or MER.[64] A further assessment by Hawksworth in 2010 concluded that China's GDP (PPP) would equal that of the US by around 2020, and that by 2030 it would be clearly the world's largest economy by any measure.[65] In a later and slightly more cautious report he assessed that China's economic progress would be "slower but equally inexorable", resulting in its GDP (MER) being larger than that of the US by 2032 if it pursued "growth-friendly policies" and if there were "no catastrophic shocks".[66]

Hawksworth was not the only analyst who became progressively more optimistic about China's prospects, at MER, for overtaking the US economy. The most famous of these more optimistic forecasts was the 2003 Goldman Sachs report mentioned in Chapter 2. It included an eye-catching graphic indicating that China's GDP (MER) would overtake Japan in 2016 to become the world's second largest economy, and that as early as 2041 it would overtake the US to become number one. The report was based on the conservative assumption that China's real growth would fall steadily from 7 to 8 percent in the 2000s to 5 to 6 percent in the 2010s, and to less than 4 percent by 2040. But it also took into account the likelihood of significant appreciation of China's currency.[67] Goldman Sachs attracted a similar amount of attention when it issued its revised forecast in 2007 that suggested China's GDP (MER) would overtake the US much earlier, by 2027. By 2050 it would be 84 percent larger than the US.[68]

Other optimistic forecasts in the mid-2000s for China overtaking the US at MER included Carsten Holz's estimate that this would occur as early as 2025.[69] Barry Naughton thought China would reach parity in about 2035.[70] Meanwhile China's currency continued to appreciate against the US dollar, which may have been a factor in Albert Keidel's prediction (also noted in Chapter 2) that China's GDP (MER) would match the US by 2035 and almost double it by 2050, and in PPP terms would overtake the US in 2020 and almost double it by 2040.[71] This drew a well-judged comment from the CIA's Stephen Voth that small differences in assumptions can result in large changes to long-term forecasts. For example, if Keidel's assumed GDP growth was reduced by half a percent and if the *yuan* converged with the dollar by 2040, then China's GDP would be only a quarter larger than the US rather than double – which implied "a different picture of the world in 2050".[72]

After the 2008 GFC, and its huge impact on most economies other than China's, there was a virtual flood of new forecasts about when China's economy might overtake the US. The wildly diverging conclusions of these forecasts raised serious questions about their credibility and usefulness. Forecasts at MER included a report by Uri Dadush and Bennett Stancil projecting that the US would give up its place as "the world's preeminent economic power" to China in 2032 (at MER), and by 2050 China would be 20 percent larger.[73] Standard Chartered Global Research in 2010 argued that the world was entering a new "super-cycle" of sustained high growth and therefore projected that China's GDP (MER) would overtake the US by 2020 and be almost double by 2030.[74] HSBC, in startling contrast, estimated that by 2050 China's GDP (MER) would still be only 10 percent greater than that of the US.[75]

In the midst of these claims and counter-claims about China's future level of global economic preponderance, several new estimates began to reflect an acceptance, at least by some experts, that China's economy was likely to overtake that of the US sooner rather than later. Conflicting views about whether China's GDP should be measured in terms of PPP or MER seemed less relevant, because even in terms of the more exacting MER measure it appeared very probable that China's ascendency was approaching and could only be upset by a major new disturbance. An article in *The Economist* in late-2010 included an interactive graphic that allowed readers to input their own assumptions about the growth of GDP (MER) for China and the US, and also about inflation and currency appreciation rates (both of which are likely to hasten the closing of the gap between China and the US). The graphic appeared to make the point quite convincingly that the date by which China could overtake the US was not particularly sensitive to changes in any of these variables. Even quite pessimistic assumptions for China and optimistic assumptions for the US, or vice versa, did not greatly affect the result. The *Economist*'s "best guess" was that China would grow at 7.75 percent and the US at 2.5 percent, inflation would average 4 percent in China and 1.5 percent in the US, and the *yuan* would appreciate by 3 percent annually. This resulted in China overtaking the US in 2019. If China grew at only 5 percent, this would only delay catch-up to 2022.[76]

Similar conclusions were reached by John Ross at Shanghai's Jiao Tong University who agreed the "central date" for China's economy to overtake the US was 2019, within a date range of 2017 to 2021.[77] Yao Yang at Peking University calculated that if China grew by 8 percent annually and the US by 3 percent, if China's inflation rate was 3.6 percent and the that of the US 2 percent (the averages for the past decade), and if the *yuan* appreciated 3 percent annually (the average for the past six years), China would become the world's largest economy by 2021.[78]

Despite the apparently emerging consensus that China would overtake the US economy by about 2020, it is interesting that the US National Intelligence Council's *Global Trends* reports continued to hold out in favour of a later date. The 2004 report stated, very conservatively, that most forecasts indicated China's GDP, in dollar terms, could be only the world's *second* largest by 2020, while the 2008 report saw this occurring even later by 2025. This was quite astonishing given that China actually became the world's second largest economy (at MER) only two years later in 2010. The 2012 report was also still quite conservative, estimating that China would probably not surpass the US until "a few years before 2030". Moreover, three of the report's four scenarios for 2030 indicated that the US economy at that time would remain larger than China's. These conservative conclusions might have been based on an assumption that China's GDP growth would drop sharply by 2020, perhaps due to it facing a "middle-income trap" and other challenges noted in the report.[79]

Economic transition

The doubts raised in *Global Trends 2030* about when China's economy might overtake the US serve to introduce another much-discussed critical uncertainty. This concerns China's ability to successfully manage transition to a more sustainable economic model that would be more suited to its long-term development needs. This qualitative question is perhaps even more important than the quantitative questions considered above about China's future rate of growth and when it may become the world's largest economy. On this issue, as on so many others, expert opinion is divided. Can China adapt its economy successfully and become even more powerful and influential? Or might failure to achieve this transition bring about the collapse, or partial collapse, about which there has been so much speculation?

China's economy has admittedly been in a state of transition for decades, and will no doubt remain so for decades to come, but two potential obstacles are increasingly debated among China experts. First, as noted in passing in previous chapters, many analysts have pointed to China's changing demography and especially to the implications for growth of the approaching peak and subsequent decline in the population of working age. Second, many economists believe that China's very high level of investment as a proportion of GDP and its correspondingly low level of domestic consumption are unsustainable and that its economic model therefore needs to change fundamentally. Underlying these issues is the theoretical concept of a "turning point", first articulated by economist Arthur Lewis in the 1950s. He believed developing countries would face this turning point when the availability of cheap rural labour began to decline and wages therefore began to rise. This would progressively undermine the viability of labour-intensive industries and require a transition to a new development model. Many theorists also argued that when per capita incomes reached a certain level this would create a debilitating "middle-income trap" as labour costs led to declining international competitiveness.

During the past decade, a great deal of attention has focused on China's demographic challenges. Much of this commentary has been pessimistic, based on UN population projections that China's working-age population will peak around 2015. Nicholas Eberstadt believed demographic pressures were likely to buffet China in a "perfect storm" causing its future growth to be "much slower" than had been assumed.[80] Similarly, Wang Feng foresaw "a looming crisis" that was "imminent and inevitable".[81] Many others accepted that in the longer term the aging of China's population would have a serious impact but were considerably more confident about the potential to offset its negative effects. Goldman Sachs noted concerns that China would "get old before it gets rich", but countered that the continued availability of surplus labour from the agricultural sector and the rapid growth of human capital through education would delay serious impacts arising from labour shortage. It considered that, by the time China became an aged society in 2027, it would already have attained developed country status.[82] Robert England also noted that China would soon begin to age rapidly – by 2025, 13 percent would be over 65, and 23 percent by 2050. Nevertheless, he believed this was unlikely to be "a significant dampener" until 2020 due again to surplus rural labour. It would not prevent China's emergence as an economic superpower.[83]

The prevailing opinion appeared be that demographic factors, although important, need not be a major show-stopper for China. They did, however, draw attention to a second and related obstacle that was of even greater concern. This was the perceived need for China's overall economic model or strategy to be changed before negative demographic factors were reinforced by other unsustainable aspects of the economy. In particular, it was thought that the economy would need to rely less on the stimulus provided by investment and exports, and much more on substituting domestic consumption as the main growth driver and means to address income inequality. Many, including apparently the Chinese government, agreed with this assessment. Premier Wen warned of underlying problems in an economy that was becoming increasingly "unstable, unbalanced, uncoordinated, and unsustainable". The World Bank and IMF were among the first to highlight this issue. They assessed that a growth pattern based on massive investment in industrial production was lifting already high investment to "unsustainable levels" and called for a rebalancing of the economy by reducing subsidies to industry and investment and developing the services sector.[84] Nicholas Lardy expanded on these concerns. He noted that China's leadership had agreed in December 2004 to "fundamentally alter the country's growth strategy" toward primary dependence on domestic consumption. While this was "laudable", he believed subsequent policy initiatives had been too modest to change the underlying growth dynamic. Household consumption was only 38 percent of GDP, the lowest for any major economy. Investment had reached 50 percent of GDP, leading to inefficiency and overcapacity.[85]

Some economists were relatively optimistic about China's ability to negotiate the transition. Ross Garnaut believed the turning point would not be as sharp as it had been in other East Asian economies, because China's greater size, diversity and internal barriers meant the impact of labour scarcity would be felt more gradually.[86] After the GFC, Garnaut considered that "large and immediate structural change" was required, but China could continue to grow rapidly if it expanded domestic demand.[87] Referring to the alleged middle-income trap, Arvind Subramanian pointed out that Japan's decline had come only after it had succeeded in catching up with the per capita incomes of the advanced economies – whereas China's per capita income was still only about one quarter of the United States.[88] CASS also appeared to be relatively sanguine. It expected continued rapid urbanisation would "keep the economy on the fast track" for 15 to 20 years. It predicted that China's urban population would reach 52 percent by 2015 and 65 percent by 2030.[89] McKinsey Global Institute estimated that one billion people would live in China's cities by 2030.[90] This would create a massive new middle class whose expenditures would boost consumption. The World Bank estimated that China's middle class would increase from 56 million in 2007 (13 percent of the global middle class) to 361 million in 2030 (38 percent of the global middle class).[91]

These positive considerations notwithstanding, most foreign experts continued to emphasise the need for China to transform its economic model, and to express concern about its ability and willingness to do this. Despite the objectives set in the Eleventh Five Year Plan (2006–2010) and repeated in the Twelfth Five Year Plan (2011–2015), capital investment remained very high as a proportion of China's GDP while household consumption languished. Revisiting this issue in 2012, Nicholas Lardy commented that it was "far from clear" that China would make the "concerted and sustained effort" required for economic rebalancing. He believed that the beneficiaries of unbalanced growth had acquired disproportionate influence in economic policy and had to date succeeded in blocking reforms.[92]

Among experts who were especially concerned about China's transition prospects, one of the most pessimistic has been economist Michael Pettis at Peking University. Noting that China's economic model transfers low interest savings from households to the corporate sector for mega-investments, he argued that such investment – especially in infrastructure and real estate – was "increasingly wasteful". It needed to be replaced as the main engine of growth by much greater household consumption, which at only 35 percent of GDP was "far less than any other country". Pettis differed from most other experts, however, in claiming that all scenarios for rebalancing the economy would require growth to fall substantially during the transition period. Chinese economists, he said, were projecting that annual GDP growth would soon fall to 6 or 7 percent, but he maintained that the sharp reduction of investment that would be a necessary precondition for successful rebalancing would cause growth to fall even more dramatically. He predicted that by 2013–14 GDP would "slow sharply" and by 2015–16 "predictions of a sustained period of growth rates at 3 percent or lower will no longer seem outlandish".[93] Pettis didn't think this would necessarily cause China to "collapse". It would "adjust", even though this would occur "neither easily nor quickly".[94] Noting that the World Bank had already reclassified China as an upper middle income country, he observed that many countries had failed to transition at this stage to a new growth model. Japan, for example, had experienced "lost decades" of very low growth.[95]

A substantial additional contribution to the discussion of China's economic transition prospects was made by Barry Naughton. He gave full credit to the Hu-Wen leadership for a "staggering" array of new and beneficial welfare initiatives, but believed they were being "swamped" by the old growth model. Looking back over the period since reform began in 1978 he commented that many experts had predicted crisis but all had been proved wrong. He therefore considered it would be foolhardy to predict a crisis in the immediate future. However, he also emphasised that China faced a new and complex mix of economic challenges. He had "a fairly high degree of confidence" that by around 2015 it would be transitioning to "a slower but higher quality growth path", but doubted this would be easy.[96]

By 2011, Naughton saw worrying indications that China was adopting a risky new economic strategy that sought to encourage both greater domestic consumption *and* increased state investment at the same time. He described this as a "new state activism", marked by "a dramatic increase both in the degree of direct government intervention in the economy and in the volume of resources going into that initiative". This "qualitatively different model of government-economy interaction" had not, he believed, received the attention it deserved. China's policymakers now envisaged "a vital, ongoing, and vigorously deployed role for the state". He believed this placed at risk previous hard-won progress with market-oriented reform. He ventured no predictions but considered that a "crucial and unknowable factor" would be the ability of policymakers to later scale back the economic role of the state.[97]

After the GFC, Naughton elaborated, China's policymakers had apparently concluded that robust intervention in the economy should be the new norm. In this new era "vigorous, coordinated, top-down measures to shape the economy" were seem as a sign of the superiority of the Chinese system. He cited Wen

Jiabao's support for making use of "both market mechanisms and … the advantages of the socialist system, which are efficient decision-making, a powerful organization, and the concentrated power to accomplish big things". Naughton emphasised that market mechanisms in China remained strong, but they were now to be accompanied and reinforced by an "acceleration of government activism". He saw this as "an enormous gamble". If successful, China would "leap into middle-income status with a speed of transformation never seen before in human history". Huge supporting investments were being made that would drive the whole development process. Naughton judged, however, that the success of this radical new approach to economic governance was likely to be mixed.[98]

An alternative appraisal of China's transition outlook was offered in a new assessment by Nicholas Lardy. Like Naughton, he warned that China's economic distortions caused by "superelevated levels of investment and systematic suppression of consumption" were accumulating and risked "a sharp and wrenching correction". This would require the government, contrary to the direction described by Naughton, to allow the market to guide "ever larger" areas of the economy. Lardy believed, however, that such rebalancing was "eminently achievable". Investment would need to be reduced from an "extraordinarily high" 48 percent of GDP to about 38 percent. To support the government's 7.5 percent GDP growth target, this would require the share of consumption in GDP to grow by a similar percentage. He recommended that this be done gradually over a ten-year timeframe because "unwinding these imbalances more quickly" would entail "a lurching correction in investment growth". In contrast with his earlier concerns, he now thought the "common refrain" that economic reform would be defeated by powerful vested interests was overstated. Because the social transfers and income redistribution included in rebalancing would have "immense public popularity", vested interests would "have a hard time resisting".[99]

Eswar Prasad, in 2013, was also optimistic about the prospects for China's economic transition. He agreed with Lardy that it would be difficult for vested interests to block reforms because they would be "sold as being for the benefit of the masses". He saw evidence that, although it was "far too early to declare victory", China was already witnessing "a gradual but remarkable transformation" in its growth model. Private and government consumption combined had already reached more than half of GDP and the decline in the share of private consumption had been halted. Physical capital investment and exports were no longer the dominant contributors to growth.[100]

Although Pettis, Naughton, Lardy and other experts clearly differed substantially about China's prospects for rebalancing its economy, they all accepted that the government's ability – or inability – to manage the transition successfully would have a major if not decisive influence on China's future outlook.

Economic futures

To summarise, the discussion above indicates that expert forecasts about critical uncertainties in China's economic future varied enormously on the quantitative issues, although the few who believed its economy might decline dramatically were outnumbered by those who were somewhat more optimistic.[101] A main-stream view emerged that lower but still healthy economic growth was likely to continue, albeit with possible short-term disruptions. A similar majority also now concedes that China is likely, probably within one or at most two decades, to overtake the US economy. On the qualitative question of successful or unsuccessful transition there were also different views, combined with much uncertainty. Many experts increasingly drew attention to major problems with the *quality* of China's growth. Some noted that failed transition to a more sustainable model could significantly alter earlier quantitative forecasts. Generally speaking, however, judgements in this domain were not as polarised as they were in the political domain.

Two principal difficulties for China's economy were raised, the first of which was the possibility that its aging population could undermine growth by reducing its comparative economic advantage. This argument attracted much attention, but many experts considered that a detrimental turning point could be delayed and/or offset by reliance on the greater productivity likely to result from a better-educated workforce and more-advanced technology. Experts generally considered it likely, however, that more-intractable obstacles would emerge unless China moved to an economy that relied less on continued very high levels of investment and more on domestic consumption. Despite clear statements of government support for such a transition, most experts agreed that there has not yet been major progress in this direction. Several expressed concern about the potential for strong vested interests, including from within the Party and the government, to continue to frustrate such plans; although others were more hopeful. Experts have also raised questions about the optimum configuration of a revised economic model, including to what extent high investment and state intervention in the economy is a liability. Lack of progress towards economic transition and slower economic growth has injected a degree of scepticism into the debate about China's economic prospects.

Many of the issues in the economic domain have naturally favoured the use of quantitative forecasting techniques, including econometrics and other forms of predictive modelling. Huge databases and complex formulas are employed, resulting in a multitude of estimates for China's future GDP long-term growth rate. The repeated shortcomings of such economic modelling were amply demonstrated by, for example, the consistent underestimation of China's likely growth – even by those who were the most optimistic. No matter how sophisticated the models, their reliability depended critically on the accuracy of their many assumptions and on the extent to which they took account of a potentially huge number of variables. The record of China experts in predicting its economic progress has therefore been very mixed. In their defence, however, most economic forecasters readily conceded and even emphasised the limitations of the forecasting tools available to them. Quantitative forecasting of China's main economic variables will continue to be an imperfect and iterative process that can perhaps be improved only marginally by greater recognition of the difficulty and uncertainty of the undertaking.

Expert analysis of China's economic growth prospects could also benefit from greater attention to the powerful influence of human actors. China's state planners have by common agreement been unexpectedly successful in the past in managing its still rapidly growing economy. Their future role may be either the key to producing even better results, or an unintended liability for developing a more sustainable economy that encourages quality rather than merely quantity of growth. The pace of economic reform is still in doubt, despite the government's declared intentions.

The expert debate about when China would become the world's largest economy has also been marked by divergent forecasts and questionable assumptions. In particular, such forecasting has been seriously hampered by the surprising and even embarrassing absence of an accepted and plausible basis for comparing the size of national economies. The appeal of using PPP comparisons for this purpose seems compelling because it defies credibility to rely entirely on MERs which ignore the fact that many goods and services can be provided much more cheaply in China than in developed countries. Some experts believe that China's GDP may already have overtaken the US in PPP terms, and even if this is not the case most agree that this will certainly occur quite soon. However, the reliability of PPP measures has been discredited by the procedures used to compare domestic prices of goods and services in each country, and by major and erratic revisions of PPPs during infrequent updates. As an apparent consequence of these difficulties with PPPs, primary reliance continues to be placed on misleading comparisons based on MERs. Given the extent of this methodological confusion, it is perhaps not surprising – although certainly concerning – that there is still considerable variance in expert predictions about when China's GDP will surpass the US, although less so than a few years previously when estimates differed enormously.

Past experience has to date defied the conviction of many experts that extrapolation of previous trends was the least likely path for China's economy. It is ironic that such simple extrapolation would, with hindsight, often have been much more accurate than the results of intensive modelling. That may not always be the case if such extrapolation is invalidated by the failure of China's leadership to manage economic transition successfully. Even with slightly slower growth, however, China's still-spectacular economic development continues to present increasingly daunting challenges for its environmental management, as will be explored in the next chapter.

Notes

1 For example, they used variations of the Cobb–Douglas Production Function including $Y=AK^{a}L^{1-a}$ where A represented productivity, K capital and L labour.

2 Although difficult to verify objectively, it seems reasonable to speculate that this underestimation may to some degree have been due to persistent bias on the part of Western economists. Such bias may have stemmed especially from an often-encountered view that China's economic prospects would be seriously constrained unless it adopted further reforms that would make it conform much more closely to a free market economy. At least in the short-term, this view did not appear justified.

3 *China: Long-Term Development Issues and Options*, 1–43. Projections were based on a model that provided "a simplified representation" of 20 sectors of China's economy, including 600 variables and "many assumptions".
4 *China: Reform and the Role of the Plan in the 1990s*. These GDP estimates were developed using "a new standard projections model" that was drawn up before growth soared to 14 percent in 1992.
5 *China 2020: Development Challenges in the New Century*, (World Bank, 1997).
6 L. Alan Winters and Shahid Yusuf, *Dancing with Giants: China, India and the Global Economy* (Washington, DC: World Bank, 2007), 5–7.
7 Louis Kuijs, *China Through 2020: A Macroeconomic Scenario* (Beijing: World Bank, June 2009).
8 *China Daily*, 24 March 2011. See also Ye Xie, "China Growth Can Beat 8% for 20 Years on Reform, Says Lin", *Bloomberg* (8 January 2013).
9 Charles Wolf, Jr, Gregory Hildebrandt and Michal Kennedy, *Long-Term Economic and Military Trends, 1950–2010* (RAND, April 1989). This report raised again the enduring question of the appropriateness of using the purchasing power parity measure of GDP, which adjusted GDP figures to take into account the lower prices for domestic consumption of various goods and services in many countries. RAND thought that PPP more accurately reflected "real resource parities".
10 Charles Wolf Jnr and K. C. Yeh, *Long-Term Economic and Military Trends 1994–2015: The United States and Asia* (RAND, 1995).
11 Charles Wolf Jnr, Anil Bamezai, K. C. Yeh, and Benjamin Zycher, *Asian Economic Trends and Their Security Implications* (RAND, 2000), and Shuxun Chen and Charles Wolf, Jr, *China, the United States and the Global Economy* (RAND, 2001), 3, 91, 94, 108 and 113.
12 Wolfgang Michalski, Riel Miller, and Barrie Stevens, "China in the Twenty-first Century: An Overview of the Long-term Issues", in *China in the 21st Century: Long-Term Global Implications* (OECD, 1996). A similar but more detailed presentation of the OECD scenarios is available in Dwight H. Perkins, "China's Future: Economic and Social Development Scenarios for the Twenty-first Century", in *China in the 21st Century: Long-Term Global Implications* (OECD, 1996).
13 Richard Herd and Sean Dougherty, "Growth Prospects of China and India Compared", *The European Journal of Comparative Economics* 4, no. 1 (2007), 41 and 65–89.
14 *China's Economic Future: Challenges to US Policy*, ed. Joint Economic Committee, Congress of the United States (New York: M. E. Sharpe, 1997), x and 101–102.
15 Murray L. Weidenbaum, "Three Scenarios on the Future of China", *Society* 34, no. 4 (May 1997).
16 Gang Lin *et al.*, *China's Economy: Will the Bubble Burst?* (Washington, DC: Woodrow Wilson International Center for Scholars, 2003), 8–13.
17 Loren Brandt, Thomas G. Rawski, and Lin Gang, eds. *China's Economy: Retrospect and Prospect* (Woodrow Wilson International Center for Scholars, 2005), 56 and 60.
18 John Wong, "China's Economy", in *Understanding Contemporary China*, ed. Robert E. Gamer (Bathurst, NSW: Crawford House, 2003), 114 and 132.
19 Nicholas Lardy, "China will Keep on Growing", *The Wall Street Journal Asia* (14 June 2002).
20 Martin Wolf, "Why is China Growing so Slowly?", *Foreign Policy* (2005).
21 Fogel, *Why China is Likely to Achieve its Growth Objectives*. See also Robert W. Fogel, "$123,000,000,000,000: China's estimated economy by the year 2040. Be warned", *Foreign Policy* (January/February 2010). For criticism of Fogel's projections see Nicholas Consonery, "A $123 Trillion China? Not Likely", *Foreign Policy* (7 January 2010), and Gordon G. Chang, "China's Economy to Reach $123 Trillion?, www.forbes.com, 8 January 2010.
22 Albert Keidel, "Why China Won't Slow Down", *Foreign Policy* (2006). See also Brandt, Rawski, and Gang, *China's Economy: Retrospect and Prospect*, 55–67.

23 *Reframing China Policy Debate 2: China's Economy*, (Carnegie Endowment for International Peace, 1 December 2006). See also Rohit Talwar, *The Future of China's Economy: the Path to 2020 – Opportunities, Challenges and Uncertainties* (Global Futures and Foresight & Fast Future, 2007).
24 Dwight H. Perkins, "Have China's Economic Reforms Stalled", in *China's Political Economy*, ed. Gungwu Wang and John Wong (Singapore: World Scientific, 1998), 41.
25 Dwight H. Perkins, *The Challenges of China's Growth* (Washington, DC: American Enterprise Institute, 2007), 11–26. Perkins based his analysis on Robert Solow's production function.
26 Dwight H. Perkins and Thomas G. Rawski, "Forecasting China's Economic Growth to 2025", in *China's Great Economic Transformation*, ed. Loren Brandt (New York: Cambridge University Press, 2008).
27 Liu Guoguang, Ga Guopei and Liang Wensen, *China's Economy in 2000* (Beijing: New World Press, 1987), 57–58, 97, 106–108 and 520–536.
28 Li Shantong and Fan Zhai, *Prospects for China's Economic Development in the Next 20 Years* (Beijing: State Council, c.1999).
29 Cai Fang and Wang Meiyuan, "How Fast and How Far Can China's GDP Grow?", *China and World Economy* 5 (2002).
30 Guo Wanda, "The Impetus and Resistance of Chinese Long-Term Economic Growth", *China Development Institute Review* (January 2004).
31 Fu Jun, "Can the Growth of the Chinese Economy Be Sustained?", *Harvard China Review* (Spring 2005), 52–56.
32 Wang Yahua and Hu Angang, "Multiple Forces Driving China's Economic Development: A New Analytical Framework", *China and World Economy* 15, no. 3 (2007).
33 He Jianwu, Li Shantong, and Sandra Polaski, *China's Economic Prospects 2006–2020* (Carnegie Endowment for International Peace, April 2007).
34 *China Embraces the Market: Achievements, Constraints and Opportunities* (East Asia Analytical Unit, Department of Foreign Affairs and Trade, Australia, 1997), 45–46.
35 Ibid., 44 and 53–55. See also Harry X. Wu, "China's GDP Level and Growth Performance: Alternative Estimates and the Implications", *Review of Income and Wealth* 46, no. 4 (December 2000).
36 Hu Angang and Men Honghua, "The Rising of Modern China: Comprehensive National Power and Grand Strategy", in *Rising China and the East Asian Economy* (Seoul: International Conference by KIEP, 19–24 March 2004), 4–5, 23, 26 and 31. This paper was translated from the Chinese version published in *Strategy and Management*, No. 3, 2002. In 2013, the Chinese Academy of Sciences reportedly assessed that China would surpass the US "in an all-round way" by 2049. See "China to surpass US by 2049: report", *Xinhu*a (8 January 2013).
37 Emilio Casetti, "Power Shifts and Economic Development: When Will China Overtake the US?", *Journal of Peace Research* 40, no. 6 (November 2003), 661–675.
38 Ahya *et al.*, *India and China: New Tigers of Asia.*
39 "The World in 2026: Who will be number one?", *The Economist* (18 November 2005).
40 Carsten A. Holz, *China's Economic Growth 1978–2025: What We Know Today about China's Economic Growth Tomorrow* (Hong Kong University of Science and Technology, 2005).
41 Angus Maddison, *Chinese Economic Performance in the Long Run 960–2030 AD*, Second ed. (OECD, 2007), 11, 14 and 20.
42 Angus Maddison and Harry X. Wu, "Measuring China's Economic Performance", *World Economics* 9, no. 2 (April–June 2008), 18–19 and 35.
43 Yuri Dikhanov and Eric V. Swanson, "Maddison and Wu: 'Measuring China's Economic Performance'", *World Economics* 11, no. 1 (January–March 2010), 199–203.

44 Albert Keidel, "The Limits of a Smaller, Poorer China", *Financial Times* (14 November 2007).
45 "A Less Fiery Dragon?", *The Economist* (29 November 2007).
46 Keith Bradsher, "A Revisionist Tale: Why a Poor China Seems Richer", *New York Times* (21 December 2007). For a more detailed discussion of the difficulties involved in the World Bank's International Comparison Program see Angus Deaton and Alan Heston. "Understanding PPPs and PPP-Based National Accounts". *American Economic Journal: Macroeconomics* 2, no. 4 (October 2010), and Robert C. Feenstra, Hong Ma, J. Peter Neary and D. S. Prasada Rao, *Who Shrunk China?: Puzzles in the Measurement of Real GDP* (University of Oxford, September 2011).
47 Mark Weisbrot, "2016: when China overtakes the US", *Guardian* (27 April 2011). However, an IMF spokesman was quoted as saying that PPPs were "not the most appropriate measure for comparing the relative size of countries to the global economy", and that the IMF believed MERs provided "a more relevant comparison". Under this measure, the US would still be 70 percent larger in 2016.
48 Yao Yang, "When will China's economy overtake America's", *China Daily* (4 June 2011).
49 Tiernan Ray, "Conference Board: China's GDP to Surpass US?", barrons.com, 10 November, 2010. However, The Conference Board's PPP converters were not the same as those used by the World Bank and IMF.
50 Wang Xiaotian and Zhang Qi, "China set for No. 1 spot by 2030", *China Daily* (25 March 2011).
51 Arvind Subramanian, "Is China Already Number One?", (Peterson Institute for International Economics, 13 January 2011). He based his findings on new estimates in the Penn World Tables that adjusted the World Bank's 40 percent reduction of China's GDP (PPP).
52 Arvind Subramanian, "The Inevitable Superpower: Why China's Dominance is a Sure Thing", *Foreign Affairs* 90, no. 5 (September/October 2011), and Arvind Subramanian, *Eclipse: Living in the Shadow of China's Economic Dominance* (Washington, DC: Peterson Institute for International Economics, September 2011), especially Chapter 5.
53 Subramanian, *Eclipse*, 109. This prediction was disputed in a review by Martin Wolf in the *Financial Times* (16 October 2011).
54 See for example Michalski, Miller, and Stevens, "China in the Twenty-first Century: An Overview of the Long-term Issues", and Weidenbaum, "Three Scenarios on the Future of China".
55 Fu Jun, "Can the Growth of the Chinese Economy Be Sustained?"
56 Richard N. Cooper, "Whither China?", *Japan Center for Economic Research Bulletin* (September 2005). Cooper believed PPP-based comparisons were problematic because China was "tied to the world economy at market prices", and because PPP rates were based on "fragile" and "arbitrary" price comparisons. He considered that, for geo-political and geo-economic purposes, MERs were more relevant. Susan Shirk agreed, considering PPPs did not "make sense" for comparing the economic and military power of different countries; see Shirk, *China, Fragile Superpower*, 273.
57 "The World in 2026: Who will be number one?", *The Economist* (18 November 2005).
58 Winters and Yusuf, *Dancing with Giants*, 5–7. See also Feenstra *et al.*, *Who Shrunk China*.
59 John Berry, "China's Booming Economy Will Never Surpass US", *Bloomberg*, (28 December 2005).
60 John A. Tatom, "Will China Surpass the United States?: The dangers of the game of extrapolation", *The International Economy* (2007), 38–41.

61 Economist Intelligence Unit, "Foresight 2020: Economic, industry and corporate trends".
62 Derek Scissors, *10 China Myths for the New Decade* (The Heritage Foundation, 28 January 2010).
63 John Hawksworth, *The World in 2050: How Big Will the Major Emerging Market Economies Get and How can the OECD Compete?* (PricewaterhouseCoopers, 2006), 3–7 and 21. He agreed with those who considered that PPPs gave a better idea of the long-term comparative size of national economies.
64 John Hawksworth and Gordon Cookson, *The World in 2050: Beyond the BRICs, a Broader Look at Emerging Market Growth Prospects* (PricewaterhouseCoopers, 2008), 2, 12 and 15–16. Hawksworth also drew attention to the World Bank's revision of the price basis for its estimate of China's GDP (PPP), which meant that instead of being three-quarters of the size of the US economy in 2008 it was now only estimated to be one half.
65 John Hawksworth, *Convergence, Catch-Up and Overtaking: How the balance of world economic power is shifting* (PricewaterhouseCoopers, January 2010).
66 John Hawksworth and Anmol Tiwar, *The World in 2050: The accelerating shift of global economic power – challenges and opportunities* (PriceWaterhouseCoopers, January 2011), 3, 6, 12, 18 and 20. Hawksworth assumed China's average annual growth rate to 2050 would be 5.9 percent compared with 2.4 percent for the US. He emphasised that it was crucial to factor in inflation and exchange rate appreciation in emerging economies, which would cause their PPPs to converge more quickly over time with MERs.
67 Wilson and Purushothaman, *Dreaming with BRICs: the Path to 2050*. The report stressed that any long-term projection was "subject to a great deal of uncertainty" because the further ahead you look "the more uncertain things become".
68 Dominic Wilson and Anna Stupnytska, *The N-11: More Than an Acronym* (Goldman Sachs, 28 March 2007), 8 and 18. See also: Jim O'Neil and Anna Stupnytska, *The Long-Term Outlook for the BRICs and N-11 Post Crisis* (Goldman Sachs, 4 December 2009).
69 Holz, *China's Economic Growth 1978–2025*.
70 *Reframing China Policy Debate 2: China's Economy*.
71 Keidel, *China's Economic Rise – Fact and Fiction*.
72 Michael D. Swaine, *China's Economic Rise – Fact and Fiction (Transcript)* (Carnegie Endowment for International Peace, 9 July 2008).
73 Uri Dadush and Bennett Stancil, *The World Order in 2050* (Carnegie Endowment for International Peace, April 2010).
74 Standard Chartered Global Research, *The Super-Cycle Report* (Standard Chartered Bank, 15 November 2010), 5, 21 and 30. Standard Chartered declined to use PPP measures because they could be "difficult to interpret" and consistent PPP weights were "difficult to estimate across time". It assumed substantial appreciation of the *yuan* and average annual growth of 6.9 percent to 2030.
75 Karen Ward, *The World in 2050: Quantifying the shift in the global economy* (HSBC Global Research, January 2011), 3 and 22.
76 *The Economist* (16 December 2010) and (10 January 2011). In a further article on 31 December 2011 *The Economist* updated its prediction for when China would overtake the US to 2018. Its blog later accepted a bet from Michael Pettis that this will not happen, with Pettis basing his own prediction on China achieving barely 3 percent annual growth in the decade to 2020.
77 John Ross, "Even Optimists Underestimated China's Growth", china.org.cn, 30 January 2011, and John Ross, "The Central Date for China to Overtake the US at Market Exchange Rates is 2019", http://ablog.typepad.com/keytrendsinglobalisation/2011/02/the-central-date-for-china.html, accessed 17 November 2011.
78 Yao Yang, 4 June 2011, op. cit.

79 *Global Trends 2030: Alternative Worlds*, iv-vii, 44–49 and 112.
80 Nicholas Eberstadt, *The demographic risks to China's long-term economic outlook* (Swiss Re – Centre for Global Dialogue, 24 January 2011).
81 Wang Feng, "China's Population Destiny: The Looming Crisis", *Current History* (September 2010). See also Feng Wang, "The Future of a Demographic Overachiever: Long-Term Implications of the Demographic Transition in China", *Population and Development Review* 37 (Supplement) (2011).
82 Helen Hong Qiao, *Will China Grow Old Before Getting Rich?* (Goldman Sachs, 14 February 2006).
83 Robert Stowe England, *Aging China: The Demographic Challenge to China's Economic Prospects* (Westport, CT: Praeger, 2005), 117–119.
84 Louis Kuijs and Wang Tao, *China's Pattern of Growth: Moving to Sustainability and Reducing Inequality* (Washington, DC: World Bank, 2005). See also Shahid Yusuf and Kaoru Nabeshima, *China's Development Priorities* (World Bank, 2006); Chi Lo, "China's Aging Problem: Limited options, ominous risk", *The International Economy* (Fall 2006), 38, 39 and 63; Cai Fang and Wang Meiyuan, "Challenge Facing China's Economic Growth in its Aging but not Affluent Era", *China and World Economy* 14, no. 5 (2006); and Rod Tyers and Jane Golley, *China's Growth to 2030: the Roles of Demographic Change and Investment Premia* (Canberra: College of Business and Economics, Australian National University, 2006).
85 Nicholas Lardy, *China: Toward a Consumption-Driven Growth Path* (Washington, DC: Peterson Institute for International Economics, October 2006). See also Nicholas Lardy, "China: Rebalancing Economic Growth", in *The China Balance Sheet in 2007 and Beyond*, ed. C. Fred Bergsten *et al.*, and Zheng Jinghai, Arne Bigsten, and Hu Angang, "Can China's Growth be Sustained: A Productivity Perspective?", *World Development* 37, no. 4 (April 2009), 877–882.
86 Ross Garnaut, "The turning point in China's economic development", in Ross Garnaut and Ligang Song, eds. *The Turning Point in China's Economic Development* (Australian National University E Press, 2006), 1–5. See also Ross Garnaut and Yiping Huang, "Continued rapid growth and the turning point in China's development", in ibid., 27–34.
87 Ross Garnaut, "China's place in a world in crisis", in *China's New Place in a World in Crisis: Economic, Geopolitical and Environmental Dimensions* ed. Ross Garnaut, Ligang Song, and Wing Thye Woo (Australian National University E Press, 2009), 1–8.
88 Subramanian, *Eclipse: Living in the Shadow of China's Economic Dominance*, Chapter 7.
89 "Urbanization expected to fuel economy", *China Daily* (30 July 2010).
90 Jonathan Woetzel, Lenny Mendonca, Janamitra Devan, Stefano Jordan, Xiujun Li, Alexander Maasry, Geoff Tsen and Flora Yu, *Preparing for China's Urban Billion* (McKinsey, March 2008 and March 2009).
91 Maurizio Bussolo, Rafael de Hoyes, Denis Medvedev, and Dominique van der Mensbrugghe, *Global Growth and Redistribution: Are China and Japan Reshaping the World?* (World Bank, November 2007), 21.
92 Nicholas R. Lardy, *Sustaining China's Economic Growth After the Global Financial Crisis* (Washington, DC: Peterson Institute for International Economics, 2012), 3.
93 Michael Pettis, "China's Economy is Headed for a Slowdown", *Wall Street Journal* (10 August 2011); Michael Pettis, "Some Predictions for the Rest of the Decade", mpettis.com, 28 August 2011; and Michael Pettis, *The Contentious Debate Over China's Economic Transition* (Carnegie Endowment for International Peace, 25 March 2011). Similar views were expressed in George Magnus, *Asia's fading economic miracle* (Centre for European Reform, January 2013), 1 and 11–12. Another study for the US Federal Reserve projected GDP growth of around 6 percent by 2030, but a combination of "less optimistic, but still reasonable assumptions" could

reduce growth to 1.5 percent: see Jane Haltmaier, *Challenges for the Future of Chinese Economic Growth* (Board of Governors of the Federal Reserve System, January 2013).

94 Michael Pettis, "China is misread by bulls and bears alike", *Financial Times* (26 February 2010).

95 Robert Zoellick, remarks to a conference on *China's Challenges for 2030*, (Beijing, 3 September 2011). See also, "Beware the middle-income trap: China's roaring growth cannot last indefinitely", *The Economist* (23 June 2011).

96 Barry Naughton, "Economic Growth: From High-Speed to High-Quality", in *China Today, China Tomorrow: Domestic Politics, Economy, and Society*, ed. Joseph Fewsmith (Lanham, MD: Rowman & Littlefield, 2010), 71–88.

97 Barry Naughton, "China's Economic Policy Today: The New State Activism", *Eurasian Geography and Economics* 52, no. 3 (2011), 313–320.

98 Ibid., 320–328.

99 Nicholas Lardy and Nicholas Borst, *A Blueprint for Rebalancing the Chinese Economy* (Washington, DC: Peterson Institute for International Economics, February 2013).

100 Eswar Prasad, "Beijing's Steady Progress Toward Rebalancing", *Wall Street Journal* (19 February 2013). For similar optimism see also Yiping Huang, Jian Chang, and Steven Lingxiu Yang, *China: Beyond the Miracle* (Barclays, March 2013), Foreword.

101 The optimists often included a proviso that growth would need to be accompanied by further reforms, especially free market reforms.

Bibliography

China 2020: Development Challenges in the New Century. World Bank, 1997.

England, Robert Stowe. *Aging China: The Demographic Challenge to China's Economic Prospects*. Westport, CT: Praeger, 2005.

Garnaut, Ross, Jane Golley and Ligang Song, eds. *China: The Next Twenty Years of Reform and Development*. Canberra: Australian National University E Press, 2010.

Garnaut, Ross, Ligang Song and Wing Thye Woo, eds. *China's New Place in a World in Crisis: Economic, Geopolitical and Environmental Dimensions*. Australian National University E Press, 2009.

Hawksworth, John and Anmol Tiwar. *The World in 2050: The Accelerating Shift of Global Economic Power – Challenges and Opportunities*. PriceWaterhouseCoopers, January 2011.

He Jianwu, Li Shantong and Sandra Polaski. *China's Economic Prospects 2006–2020*. Carnegie Endowment for International Peace, April 2007.

Holz, Carsten A. *China's Economic Growth 1978–2025: What We Know Today About China's Economic Growth Tomorrow*. Hong Kong University of Science and Technology, 2005.

Lardy, Nicholas. *Sustaining China's Economic Growth after the Global Financial Crisis*. Washington, DC: Peterson Institute for International Economics, 2012.

Lardy, Nicholas and Nicholas Borst. *A Blueprint for Rebalancing the Chinese Economy*. Washington, DC: Peterson Institute for International Economics, February 2013.

Li Shantong and Fan Zhai. *Prospects for China's Economic Development in the Next 20 Years*. Beijing: State Council of China, c.1999.

Lin, Gang, Charles Wolf Jnr, Thomas G. Rawski and Deborah S. Davis. *China's Economy: Will the Bubble Burst?* Washington, DC: Woodrow Wilson International Center for Scholars, 2003.

Maddison, Angus. *Chinese Economic Performance in the Long Run 960–2030 AD*. Second ed: OECD, 2007.

Naughton, Barry. "China's Economic Policy Today: The New State Activism". *Eurasian Geography and Economics* 52, no. 3 (2011).

Naughton, Barry. "Economic Growth: From High-Speed to High-Quality". In *China Today, China Tomorrow: Domestic Politics, Economy, and Society*, edited by Joseph Fewsmith. Lanham, MD: Rowman & Littlefield, 2010.

Perkins, Dwight H. *The Challenges of China's Growth*. Washington, DC: American Enterprise Institute, 2007.

Perkins, Dwight H. and Thomas G. Rawski. "Forecasting China's Economic Growth to 2025". In *China's Great Economic Transformation*, edited by Loren Brandt. New York: Cambridge University Press, 2008.

Pettis, Michael. *The Contentious Debate over China's Economic Transition*. Carnegie Endowment for International Peace, 25 March 2011.

Subramanian, Arvind. *Eclipse: Living in the Shadow of China's Economic Dominance*. Washington, DC: Peterson Institute for International Economics, September 2011.

6 Environmental uncertainties

A comprehensive analysis of China's environmental trends is beyond the scope of this study. It would encompass many subject areas and require analysis of expert views about numerous uncertainties. These would, for example, include China's ability to meet its future energy requirements, the availability of natural resources, and the likely effects of pollutants on China's environment and on the health of its citizens. The main focus of this chapter will therefore be on examining the range of expert opinion about the most consequential issue in this domain, namely the likely impact of China's still massive resource and energy consumption on its future carbon emissions and on climate change. This critical uncertainty can be expressed in overall terms simply as: *What is the outlook for China's environmental management and sustainability?* From both a national and global perspective, uncertainties in this area may well be the most critical of the uncertainties in all domains.

Within this reduced subject area, primary consideration will be given to two specific critical uncertainties. The first, which has received a great deal of attention in expert analysis, approaches the question from the viewpoint of governance. It concerns both the willingness and the ability of China's political and administrative systems, at both central and local levels, to formulate and implement measures that will rein in GHG emissions. This critical uncertainty can be expressed as: *How successful, from a governance perspective, will China be in managing the negative impacts of economic growth on its environment and on global climate change?* The focus of this analysis will be on the strategies and performance of key actors and decision-makers in devising and implementing appropriate counter-measures. This includes the ability of the government to withstand the powerful forces and vested interests that drive continued expansion of resource and energy consumption and resist the imposition of measures to control resultant pollution and emissions.

The second – more strategic and more quantitative – critical uncertainty, which has begun to receive more attention, can be expressed as: *Will China's efforts to achieve environmental sustainability, by reducing its energy demand and carbon emissions, be overwhelmed (or assisted) by continued economic growth?* This critical uncertainty focuses essentially on calculating the net quantitative effect of the conflict between two opposing forces – the development-driven growth of emissions on the one hand and the success of measures to contain them on the other. This question has usually been addressed by experts through various forms of scenario-based statistical modelling.

Managing environmental impacts

With regard to the first of these two critical uncertainties about the quality of environmental governance, many experts were somewhat slow to see this as an important issue. The most prominent Western scholar to begin drawing attention to China's worrying environmental direction was Vaclav Smil. His assessment was notably pessimistic, both about the problems and about the measures being adopted by the government. He documented alarming trends in deforestation, erosion, desertification, loss of arable land, declining water resources, air pollution, and species extinction. He was encouraged by the beginnings of a bureaucratic response but also highlighted "fundamental limitations". The implications of failure were "grim", but he thought this was "at least as likely" as success. He concluded that the most fundamental check on China's future would be its "staggering mistreatment" of the environment.[1] As time went by, Smil's pessimism deepened. He forecast, with impressive anticipation, that even if China's energy consumption stabilised in the 1990s, it would expand again in the subsequent decade. China would become "the planet's leading environmental transgressor".[2] He calculated that even with aggressive environmental management it would become the world's largest GHG emitter between 2010 and 2025.[3] A spark of optimism was evident, however, in his comment that there was "no preordained progression" in China's environmental trends because "human futures, while not infinitely alterable, are amazingly malleable".[4]

Among Chinese authors, Guangdong-based academic He Bochuan was also notably pessimistic about the ability of China's leaders to manage its environmental challenges. His book *China on the Edge* was a best seller within China in the 1980s. He considered there were "no effective or feasible measures" available to counteract key environmental trends in China. It had gone "past the point of no return" and the latent environmental crisis might explode.[5]

By the mid-2000s, however, a groundswell of concern about China's environmental mismanagement had developed among many experts and commentators, which led them often to contemplate the possibility of collapse. The "terrible price" paid by China's environment in exchange for its impressive growth was dramatically documented by Elizabeth Economy in her book *The River Runs Black*. A "long, slow process" had commenced to introduce environmental protection measures, but this was not backed by sufficient empowerment of environmental agencies. She noted the potential for political discontent about environmental issues to "bring the country to its knees economically". Economy considered three scenarios: *China Goes Green*, with greater environmental investment and the CCP recast as a new political party in a 2013 presidential election; *Inertia Sets In*, with ineffective management of pollution and climate change; and *Environmental Meltdown*, including possible internal collapse and civil war.[6]

Meanwhile China's outspoken Deputy Environment Minister Pan Yue attracted international attention with his daring statement that China's economic miracle would "end soon" because the environment could not keep pace.[7] Peter Aldous warned that China's "skyrocketing" emissions were threatening global climate disaster, but deduced that this did not "loom large" in the thinking of China's leaders.[8] Jianguo Liu and Jared Diamond, noting the competition between economy and environment in China, seemed pessimistic about "which horse would win the race".[9] Other experts expressed growing concern about an emerging water crisis, which might have political consequences.[10] Peter Engardio believed "runaway environmental degradation" was the greatest challenge facing China, but counter-measures were "at best at a standstill".[11] A major assessment by international experts, headed by Peter Ho, noted that China's authorities were "painfully aware" of the severity of environmental problems but an "increasingly obsolete" Chinese state seemed incapable of meeting the challenge. Like other countries, it appeared to favour only "incremental change", which might eventually trigger self-destruction.[12]

Scepticism among experts about China's motivation and ability to manage pressing environmental issues rose to new levels in the early-2000s when its accelerating economic growth led to a breakout in energy consumption. From 2001 to 2005, GDP growth surged dramatically and energy consumption, led by a booming heavy industry sector, increased at an equal or even faster rate. This marked the end of a period of some two decades in which China's energy consumption had, very commendably, expanded at only about half the rate of GDP growth. The Aspen Strategy Group noted "growing awareness" in China of this problem, but believed it was "not making adequate investments – or using adequate technology". They concluded that without international assistance for China, catastrophic global warming would be unavoidable.[13]

China's worrying new energy consumption and carbon emissions trends were analysed by Daniel Rosen and Trevor Houser. They observed that between 1978 and 2000 China's economy grew annually by 9 percent while its energy demand grew by only 4 percent, but demand growth subsequently surged to as high as 13 percent in the 2000s. They believed China's leaders were well aware of this problem and had "lofty ambitions" to limit energy consumption, but progress was slow because "the same vested interests responsible for the current economic trajectory frustrate efforts to change course".[14]

From the mid-2000s, the Chinese government began to initiate strong countermeasures to limit energy usage and carbon emissions. But, in the period leading up to the pivotal 2009 climate change summit in Copenhagen, experts continued to question the effectiveness of these countermeasures and the extent of the government's commitment. The World Bank was among the first to voice its mounting concern. In a major new report in 2007 it considered that China's "already enormous" energy consumption might grow to a level that would be "difficult to satisfy". Even "more alarming", the authors emphasised that the "window of opportunity" to control energy consumption was closing rapidly due to the "huge current growth of investments in existing rather than new and advanced energy technologies". International Energy Agency (IEA) forecasts had indicated that China could be responsible for one-quarter of the growth in world CO_2 emissions between 2002 and 2020. Crucial reforms had "stalled", policy making was "piecemeal and uncoordinated", and implementation was "weak". The report stressed in a tone bordering on desperation that the "time for action has arrived". But it added that urgency was "the characteristic lacking" in China's policies. These blunt observations were softened only marginally by the Bank's alleged optimism that with stronger action China could transform itself "from a laggard to a global leader" in the deployment of new energy technologies.[15]

Elizabeth Economy also continued to be disturbed by China's perceived inaction. She raised the spectre of "Shanghai under water, Tibetan glaciers disappearing, crop yields in precipitous decline, epidemics flaring"; but believed China's leaders paid "about as much attention" to climate change as did George W. Bush. They showed "little inclination to move aggressively". China's GHG emissions had tripled over the last 30 years and might triple again in the next 30 years. She conceded it would be "unfair" to say China was doing nothing. It had set bold targets for energy intensity (energy consumed per unit of GDP) and for renewables in its Eleventh Five Year Plan. But its development strategy suggested "little will change in the foreseeable future". The guiding logic was still "growth at all costs".[16] Economy's critique went further. She judged that China's environmental problems were systemic, arising from its "corrupt and undemocratic political system". Without a fundamentally different political system she thought China would not "return to global pre-eminence" but would suffer "stagnation or regression".[17]

Grave concern was expressed by US researchers Malcolm Shealy and James Dorian. Noting China's rapidly expanding energy consumption they asked: "Is the world in denial?" They projected China's coal consumption might triple between 2005 and 2025, and that its carbon emissions could also triple.[18] William Chandler called for the US and China to break their "suicide pact" on climate change. He was encouraged by Chinese sources that suggested it could "cut its current emissions growth by half through 2020 and from that level reduce absolute emissions by one-third by 2050". However, although the central government had set sound high-level policies for energy efficiency, it left implementation to "unprepared, underresourced provincial leaders".[19]

Kelly Sims Gallagher was also sceptical about the true commitment of China's leaders to environmental management. She believed that they were driven primarily by the need to meet rising energy demands, which were seen as "more pressing" than global warming. In 2006 alone, she observed, China installed almost as much new power capacity as India's entire electricity system. Moreover, many local governments "ignore or flaunt even the most basic energy-efficiency policies". China's carbon emissions had grown by 9 percent annually from 1999 to 2004. It was building 34 new ultra-supercritical coal-fired power plants, but she said it had in 2006 built three times that number of traditional plants. By using the cheapest technologies available it was "locking itself and the world" into high emissions in future decades.[20] Gallagher accepted that China's leaders expressed great concern about environmental issues and had passed many laws and regulations, but provincial and local governments failed to enforce such regulations "more often than not".[21] Paul Harris shared this view, claiming that the central government had "limited control over the vast bureaucracy outside Beijing" and had succumbed to the "strong nationwide fixation on economic growth". The future looked "bleak indeed".[22]

Nevertheless, by the late 2000s some experts began to see new measures – such as China's surprise announcement that it would aim to reduce energy intensity by 20 percent between 2005 and 2010 – as indicating a promising new determination to address environmental challenges. A report by the US Department of Energy's Lawrence Berkeley National Laboratory (LBNL) considered that this new target marked "a major shift in China's strategic thinking". It also commended a program that committed 1,000 of China's largest enterprises to meet specific energy saving targets. But LBNL still appeared sceptical about whether this would be sufficient. The 20 percent target would be difficult to achieve given the recent spike in energy intensity, which had raised the spectre of "a huge 'unexpected' disturbance" in the global energy and climate system. The target was considered "extremely ambitious" and "almost certain" to fall short.[23] LBNL thought China would be hampered by many factors, including especially by the reality that in China "making money matters" and policies that obstructed this were "particularly difficult to put into practice".

In a more optimistic vein, however, LBNL commented that it would be inadvisable to underestimate the Chinese government's ability "to achieve large goals".[24] LBNL's Mark Levine testified to a US congressional committee that one of the greatest misunderstandings between China and the United States was the failure to recognise that China during the last two decades of the twentieth century, and again more recently, had put "tremendous effort" into reducing carbon emissions through implementing "aggressive and innovative" policies.[25] Indeed China showed increasing indications that it accepted the reality of climate change and its anthropogenic origins, and that it was prepared to take serious steps. In December 2006, it issued a *National Climate Change Assessment Report* that candidly assessed the recent and likely future impacts on China – including temperature rises, falling agricultural productivity, desertification, loss of biodiversity, sea level rise, permafrost degradation, glacier melting, intensified droughts, and water shortage.[26]

Reactions from international experts were still mixed, however. Joanna Lewis commented that China's efforts "should not go unrecognized", including its closure of many small high-polluting power generation and industrial plants, its aim to produce 16 percent of primary energy from renewable and hydropower sources by 2020, and the establishment of a high-level National Leading Group on Climate Change to oversee specific targets and timetables. Even so, she thought current policies would be insufficient to achieve the required emissions reductions.[27] ZhongXiang Zhang, at the East–West Center in Honolulu, was "cautiously optimistic". He noted that China had failed to achieve its targeted energy intensity reduction in 2006, but that the government was making "concerted efforts".[28] Ross Garnaut, Frank Jotzo, and Stephen Howes at the Australian National University noted it would be "in China more than anywhere else" that the success of global climate change mitigation would be decided. They thought many mainstream forecasts underestimated the challenge China faced. It was, however, "starting" to put ambitious policies in place.[29] Researchers at the Stockholm Environment Institute (SEI) were less charitable, arguing that China's policies were at best a repackaging of existing strategies that happened to have "co-benefits" for climate change mitigation. A fundamental shift was required. Obstacles – including administrative shortcomings, conflicting central–local priorities, and rent seeking and corruption – were considered to be "deeply rooted in the Chinese political system". SEI urged China to place climate change "at the core" of its strategy. It added that this issue was a "critical uncertainty" for China and the world.[30]

From early 2009, international attention began to concentrate on the prospects for progress during the Copenhagen negotiations at the end of the year. Expectations naturally centred on China as a potential "dealmaker".[31] Jianjun Tu, at the Carnegie Endowment for International Peace, described its target to reduce energy intensity by 20 percent as "one of the most significant carbon mitigation initiatives in the world".[32] The seriousness of China's intent was indicated by its top negotiator Minister Xie Zhenhua, who commented that China had "done a lot" to reduce emissions and would work "aggressively and actively" to reduce them further.[33] US Special Envoy for Climate Change Todd Stern described China's role as close to being the "alpha and omega", given that it was "rocketing up the emissions curve". He cited modelling which indicated that, even if other countries reduced emissions by 80 percent before 2050, China's emissions alone "under business-as-usual assumptions" would lead to atmospheric CO_2 levels of 540 ppm and a 2.7-degree temperature rise. But he believed China recognised that business-as-usual was unsustainable.[34]

Some experts defended China's progress. ZhongXiang Zhang detailed its positive actions and lamented that it seemed to have become "a Christmas tree on which everybody could hang his/her complaints".[35] His view was supported by a wave of major new reports that documented China's progress and manifested an increasing degree of cautious optimism. In 2008, The Climate Group, an international NGO network, published its first edition of *China's Clean Revolution*, which found that "far from ignoring climate change" China was becoming a leader in low carbon industries. The Climate Group conceded that the jury was still out on whether it could meet its challenging targets, but it would soon be "the No. 1 manufacturer of various critical low carbon technologies" and had a "strong and comprehensive low carbon policy framework". For example, it had not only closed many small and inefficient coal-fired power plants but the new plants that replaced them "almost exclusively" used supercritical or ultra-super-critical technology.[36] A second *China's Clean Revolution* report assessed that its progress might reduce the cost of low carbon technologies internationally. Tony Blair, in his foreword, noted China was "among the world's leaders" in producing wind turbines, solar panels and biofuels, and its steel and cement plants were "among the most efficient in the world". The report also recorded that China had introduced "a number of watershed laws and programs" and considered it had "the capacity to enforce these". Achievement of its 2010 energy intensity target was in sight.[37] China's low carbon credentials were also lauded in a *China Greentech Report*, which assessed that it could soon emerge as a global innovator.[38]

This optimism contrasted with a sobering IEA assessment that it would take decades for other power sources "to push coal from its dominant position" in China.[39] However, US journalist Keith Bradsher took a more positive attitude. He considered that what was "largely missing in the hand wringing" was recognition that China had suddenly emerged as the world's leading builder of more-efficient and less-polluting coal plants that were "mastering the technology and driving down the cost". This progress was "as fast and serious as anywhere in power-generation history".[40]

Julian Wong and Andrew Light at the Center for American Progress also protested against "the common refrain from climate action naysayers" alleging that China was building two new coal-fired power plants per week, making emissions reduction commitments by other countries irrelevant. They countered that China had already "forged ahead with building its own low-carbon economy" and could become the unquestioned leader in clean energy technologies. If China achieved its 2010 energy intensity reduction target, this would save one billion tons of carbon dioxide per year, compared with the EU's targeted annual saving of 300 million tons.[41] Beijing-based Deborah Seligsohn was also upbeat. She believed China's ambitious and impressive approach was under-appreciated. She perceived that from 2007 a combination of domestic and international pressure had produced a "real consensus" in China, leading to "a fully articulated climate change policy" backed by clear and measurable targets.[42] She cited IEA estimates that 80 percent of China's coal-fired plants would be modernised by 2011, and 90 percent by 2020. China's wind generation was projected by 2020 to reach five times the target set in 2005.[43]

In the lead-up to Copenhagen, China appeared to become increasingly aware of the need to show the international community that it was responsive to pressures for it to make specific commitments to limit its carbon emissions. As early as 2006, Hu Angang had advocated shifting China's growth model to "a green development path".[44] In 2009 he proposed a new approach. He argued that the current classification of nations as developed or developing did not "reflect reality" and recommended instead that global emissions reductions should be based on dividing countries into four groups, according to their ranking in the UN's Human Development Index. This would place China in the second category of the HDI, immediately after the most developed countries, and increase its expected contribution accordingly. By 2010 it was expected to reach the first category. Hu believed Copenhagen was "our last chance to save the planet".[45] Several other Chinese proposals for capping global emissions emerged at this time, but they recommended that a global emissions budget should be based on *cumulative* per capita emissions, placing the main burden on developed countries.[46]

An early sign of the Chinese government's willingness to make new commitments came from Su Wei, a leading member of its negotiating team, who reportedly stated that China could "very easily" translate its energy intensity targets into limits on CO_2 emissions.[47] This was followed in September 2009 by Hu Jintao's undertaking at a special UN summit that China would "endeavour to cut carbon dioxide emissions per unit of GDP by a notable margin by 2020 from the 2005 level". Then on 26 November 2009 the State Council made the long-awaited announcement of a voluntary but domestically binding commitment to reduce China's carbon intensity (carbon emitted per unit of GDP) by 40 to 45 percent by 2020 compared to 2005.

As events unfolded, however, Copenhagen was to be something of a public relations disaster for China, in spite of its resolve to reduce its own emissions. Its apparent inflexibility caused it to be accused by journalists and even by some negotiators of having "wrecked the talks", although others pointed the finger at the United States and developed countries.[48] ZhongXiang Zhang speculated that China's rejection of "unilateral" 2050 emissions cuts by developed countries may have been due to fear that it would be pressured to also take on additional exacting commitments, a view that has since been supported by others.[49] Nevertheless, negative international perceptions of China's willingness to act on climate change issues persisted. Joanna Lewis pointed out that between 2010 and 2030 China was expected to account for 40 percent of all new energy-related CO_2 emissions.[50]

Continuing doubts about China's environmental management were also evident in a 2010 joint report by the United Nations Development Programme and Renmin University. It commented that China had probably promulgated more laws, policies and measures for low carbon development than any other developing country, but these would be "only partially effective" without strong measures to deal with implementation obstacles that included poor coordination, weak regional and local implementation, and insufficient monitoring and enforcement.[51] A much more positive stance was taken, however, in a report by David Held, Eva-Maria Nag, and Charles Roger. They thought that there was a "notable disjuncture" between China's reputation for "ongoing obstinacy" in international negotiations and the impressive scale of its domestic initiatives. A revolution in domestic governance had made implementation of environmental policies legally mandatory and an important factor in local officials' performance evaluations. Policy-makers had shown "great determination" and had introduced new measures "on a scale nearly unparalleled elsewhere". They pointed out that this contradicted the dire picture found in international commentary that portrayed China as an "obstinate climate laggard".[52]

As the Copenhagen debacle receded further, China's ongoing efforts to stimulate development of low carbon sectors in its economy won increasing recognition. A Pew survey pointed out that China had become the leader within the G-20 in clean energy finance and investment. It was becoming "the world's clean energy powerhouse".[53] Energy Secretary Steven Chu argued the US could no longer use China's alleged inaction on environmental issues as an excuse, rather "China's rapid action should be a motivation".[54] Deutsche Bank stated that China had developed "robust policy frameworks, including clearly defined national targets, strong incentives and integrated plans". China was "a fast adopter of clean technologies" and was the leader in renewable electric power, having doubled its installed capacity during 2009 alone. Its annual clean energy investment exceeded the US and was three times greater than that of the US as a percentage of GDP.[55] Deutsche Bank saw China becoming "a game changer" for clean technology industries globally.[56]

China's Twelfth Five Year Plan included new targets to reduce energy intensity by a further 16 percent from 2010 to 2015, and to reduce carbon intensity by 17 percent. Energy efficiency, nuclear energy and renewables, and electric vehicles were highlighted as three of the seven Strategic Emerging Industries to be allocated high investment priority under the Plan, which also referred to "step by step establishment of carbon emissions trading markets". The Climate Group concluded that China placed low carbon growth "at the heart of economic strategy in a way that no other country has done". It was likely to become "the world's fastest growing market for climate solutions" and "the world's biggest exporter of key clean technologies".[57] The 2011 and 2012 *China Greentech Reports* were also enthusiastic, declaring that China was now "a greentech leader" standing "at the center of almost every greentech market" and the world's largest clean energy investor. It had been criticised at Copenhagen but was leading by domestic action. Its Strategic Emerging Industries were targeted to absorb 8 percent of GDP by 2015 and 15 percent by 2020. The report commented that it was unimaginable that Europe, Japan or North America could build 20,000 kilometres of high-speed rail in a decade, or imitate China's surge in nuclear power, or take back China's lead in wind installations.[58]

Amid this chorus of approval there remained some dissenting voices. Not all commentators were impressed. John Lee claimed that the greening of China was "a mirage". He admitted China's new Five Year Plan contained "superficial evidence" that it was taking climate change seriously, but saw this as a "carefully crafted message" designed to impress Western markets.[59] Lee advised environmentalists not to be fooled by China's plans for a carbon tax because they were "little more than clever political theater" designed to allow it "to emit even more".[60] But it was rare for commentators to display this degree of scepticism and Lee's claims drew a sharp response from US analysts who asserted his article was "plain wrong".[61]

Even other commentators who were still very concerned about environmental trends in China were nonetheless usually prepared to concede the scale and sincerity of its efforts to offset these impacts. A report by the Economist Intelligence Unit, for example, stated that 35 percent more coal would be consumed in China in 2020 than in 2010, and carbon emitted by burning coal and other fossil fuels would rise by 43 percent. It would be greener in relative terms, but grey would "remain the dominant colour". However, the Unit accepted that China was taking aggressive measures to encourage low-carbon energy, with "some notable successes". The combined share of nuclear and renewables in energy supply would rise from 13 percent in 2010 to over 16 percent in 2020. China would become "the primary engine" driving the global green economy. Trial carbon trading schemes were underway and a carbon tax, at a relatively low rate, was also considered likely by 2015.[62]

Achieving environmental sustainability

By the early 2010s, most experts who paid close attention to China's efforts to curb energy intensity and emissions appeared on balance to be impressed by the scale of its response, at least from a policy, administrative, and governance perspective. But, in quantitative terms, this still left open the second critical uncertainty identified earlier: namely, whether its relative success in reducing its energy demand and carbon emissions might still be overwhelmed by the consequences of its continued rapid economic growth – and by its continued reliance on fossil fuels and heavy industry investment. This issue has also received considerable expert attention, although limited mainly to energy and environmental specialists. Typically they used a variety of statistical modelling and scenario techniques to attempt to project trends in energy demand, energy intensity, carbon emissions, and carbon intensity.

Initially, in the 1990s, some estimates by international agencies of China's energy and emissions trends were relatively optimistic. An extremely sanguine World Bank report in 1997 thought that China could reduce its future carbon emissions in 2020 to below existing levels, even after economic output had quadrupled. Energy intensity in 2020 was expected to fall to 62 percent of its 1995 level. With only "a few crucial adjustments", an environmentally sustainable future was considered to be "well within reach of China's current policies and resources".[63] This level of optimism was almost unique, however. The IEA was impressed by China's past success in controlling energy intensity, but much more pessimistic about the longer term trends. It estimated in 1996 that China's annual CO_2 emissions would almost double by 2010 and reach 17 percent of the world total.[64]

By the turn of the century, growing concern among experts about China's global environmental impacts led to several major bilateral and international collaborative research projects and modelling exercises. The scope of collaborative activity in this field between China and other countries has had a low public profile, but is long-standing and has undoubtedly been very influential. An early example was a study in 2000 of energy futures to 2020 by the Chinese and US Academies of Sciences and Engineering. They expected that China's carbon emissions would overtake those of the US "over the next several decades".[65] In another collaborative report, the lead was taken by China's Energy Research Institute (ERI) but with strong support from counterparts in the US and especially from LBNL. Three years of joint research led to the publication in 2004 of a report in Chinese titled *China Sustainable Energy Scenarios in 2020*. Three scenarios were developed: an *Ordinary Effort* scenario that assumed a low priority for environmental sustainability, a *Promoting Sustainability* scenario that assumed implementation of new policies outlined in China's Tenth Five Year Plan, and a *Green Growth* scenario that assumed more aggressive measures. The report found that, depending on which scenario eventuated, energy use might rise by 70 to 130 percent by 2020, while carbon emissions over the same period could rise by 50 to 100 percent, or more.[66]

The shocking reversal during the early-2000s of China's energy efficiency achievements led to new efforts by Chinese and foreign experts to reassess its prospects for achieving environmental sustainability. Perhaps the first foreign analysts to give an early warning of China's energy consumption breakout, and to examine the future implications quantitatively, were William Chandler and Holly Gwin. They noted that, although for two decades previously China had held its energy demand growth to 60 percent of GDP growth (an elasticity of 0.6), in 2002–2003 that demand had grown "perhaps one-third faster than GDP" (an elasticity of 1.3). They feared this could mark "a momentous change" in China's ability to control its GHG emissions. Its energy demand could grow "nearly four-fold by 2050". It had reached a turning point where, in the absence of counter-measures, a rapid expansion of emissions "could overwhelm domestic and international mitigation efforts".[67]

Others were much slower to pick up this new trend. In a notable example of forecasting failure by multiple agencies, most analysts missed the turnaround in China's energy efficiency performance. They failed to anticipate it and they were slow to recognise it after it happened. Both the Chinese government and the IEA had predicted annual energy demand growth would remain at 3 to 4 percent up to 2010. In fact, energy elasticity reportedly increased from less than 0.5 between 1978 and 2000 to up to 1.5 between 2001 and 2006. Energy consumption grew four times faster than predicted in most forecasts.[68] Another quickly outdated assessment from the US Department of Energy's Energy Information Administration (EIA) projected that China's share of global carbon emissions from fossil fuels would grow from 13 percent to only 18 percent in 2025 – a level that in fact was reached only a few years later.[69] A 2007 Goldman Sachs analysis brought forward the time when China's CO_2 emissions might exceed those of the US to 2015, with China's share of world emissions reaching one quarter to one third by 2030 – but both mileposts were passed much earlier.[70] Researchers at CASS noted in 2006 that the Chinese literature also contained many over-optimistic scenarios for future emissions. Actual consumption in 2005 had been higher than these Chinese scenarios had projected for 2010, while consumption in 2010 seemed likely to be close to or higher than had been projected previously for 2020.[71]

In reaction to these mistaken forecasts, a 2008 report by Mark Levine and Nathaniel Aden at LBNL constituted what might almost be called an anti-forecast. Instead of offering their own projections, they dwelt on the astonishing inaccuracy of previous estimates. They noted that a "profound change" in the level of China's emissions had clearly occurred around 2002, and asked why forecasts over a long period before and after this date had been "so greatly in error". China's energy-related carbon emissions had already surpassed those of the US in 2006, something the IEA and EIA as recently as 2004 had thought would not occur before 2025 or 2030. (In other words, China's emissions overtook the US within two or three years rather than two or three decades.) Furthermore, 55 percent of global emissions growth between 2000 and 2006 had come from China alone. China's annual emissions growth had averaged 10 percent after 2000 compared to 5 percent between 1980 and 1995. By 2006 China's share of total annual global emissions had reached 20 percent.[72]

In this report, LBNL surveyed a number of forecasts over previous years by IEA and EIA, as well as three "highly respected" Chinese forecasts (from Tsinghua, ERI and DRC). It highlighted the dramatic swings in these forecasts. The most recent estimates of future emissions were the highest, but the immediately preceding forecasts were the lowest. These miscalculations were due to the astonishing fact that the three-year increase in China's emissions from 2002 to 2005 was "equivalent to the increase projected to take place over a period of fifteen or more years!" This suggested to Levine and Aden that predicting China's energy and emissions trends in coming decades would be "precarious", because this depended greatly on China's ability to "tame its burgeoning energy demand".[73] This well-meaning advice went largely unheeded, however. It was not long before a new crop of forecasts appeared, their authors apparently undeterred by their colleagues' earlier failures. In total, from the mid-2000s at least thirteen major assessments appeared that attempted either to project energy and emissions trends forward under various scenarios to 2030 or 2050, or – following a quite different approach – to project the reductions that would be required to meet more ambitious but predetermined targets.

The first of these reports were understandably pessimistic, given the recent emissions surge. Australian researchers Peter Sheehan and Fiona Sun noted that China's energy consumption in 2001–2005 had increased by an average of 9.8 percent annually, compared with forecasts ranging from 2.6 to 4.7 percent from IEA, EIA, ERI/LBNL, and China's National Development and Reform Commission.[74] Assuming continued high GDP growth and energy intensity, Sheehan and Sun projected that China's energy use and CO_2 emissions could both increase almost sixfold from 2002 to 2030.[75] In a subsequent paper, however, they seemed impressed by China's new policies, which they thought could reduce projected energy use and emissions in 2030 by up to 35 to 40 percent. They thought that the government's new target to reduce energy intensity by 20 percent between 2005 and 2010 was "unlikely to be achieved", but believed China was now committed to a "full range of instruments" that should ensure "major reductions".[76]

Alarming projections of China's CO_2 emissions path were made by several other experts. US analysts Maximilian Auffhammer and Richard Carson projected China's likely emissions to 2015 would be larger by "a factor of two or more" than previous forecasts due to alleged "abandonment of energy efficiency programs".[77] The World Bank's Zmarak Shalizi calculated that, under a business-as-usual scenario, China's energy use and CO_2 emissions might more than double by 2020 and almost double again by 2050.[78] The EIA's 2006 forecast indicated China's percentage share of both global energy consumption and global CO_2 emissions would almost double between 2003 and 2030.[79] The IEA's 2007 forecast estimated that by 2010 China would be the world's largest energy consumer, and over the period to 2030 it would install "more new electricity generating capacity than exists in the United States today". It would also account for 42 percent of the growth in global CO_2 emissions to 2030.[80] In a *Science* magazine article, Chinese and US researchers raised a hypothetical scenario in which China's carbon intensity kept pace with a GDP growth rate of 7 percent, causing its emissions to equal current global emissions by 2030.[81] An ominous report by Geoffrey Blanford, Richard Richels and Thomas Rutherford at Harvard University warned that China's emissions might soon cause international goals to stabilise atmospheric CO_2 levels to "become unattainable". China's emissions since 2000 had grown annually by over 10 percent, compared with 2.8 percent in the 1990s.[82]

A three-year study published in 2009 by the UK's Tyndall Centre for Climate Change Research appeared more hopeful. However, it was not based on the usual approach of projecting expected emissions growth under various assumptions, but on identifying scenarios with different premises that, if achieved, would all meet a *predetermined* global target of stabilising atmospheric CO_2 at 450 ppm. This was the first of several such "normative" forecasts, which concentrated only on the measures that would be needed to achieve their predetermined targets. Significantly, the study indicated China's emissions would need to peak between 2020 and 2030, with early large-scale action being critical to success. The scenarios indicated China's carbon emissions would rise by 24 to 72 percent between 2005 and 2020 depending on the size of its emissions budget, but decline thereafter by 15 to 70 percent by 2050. This was considered "challenging but achievable".[83]

Other such studies followed Tyndall's example in taking an optimistic view of the possibilities for curbing China's carbon emissions significantly if its government acted quickly and strongly. A report by McKinsey and Company, which took a similar normative approach, examined over 200 efficiency and abatement technologies to assess their maximum potential to assist China to reduce its emissions. A *Baseline* scenario concluded that for each of the five year periods to 2030 China could reduce its energy intensity by 17 to 18 percent. Even so, its absolute emissions would more than double over that period. However, an alternative *Abatement* scenario, in which China fully deployed the available technologies, could reduce emissions in 2030 to nearly half of the *Baseline*, or only 10 percent higher than emissions in 2005. In this scenario, coal's share in overall power supply would reduce from over 80 percent to as low as 34 percent. McKinsey tempered this startlingly optimistic projection by acknowledging that it would require no less than a "green revolution" and that any delay in seizing the window of opportunity would greatly reduce the potential emissions savings.[84]

Two other major studies, both from the SEI, emerged prior to Copenhagen. They recommended very aggressive and probably unrealistic emissions reductions. A joint report by SEI and leading Chinese economists contained emissions scenarios ranging from business-as-usual to a deep reduction scenario that would ensure a global temperature increase of no more than 2 degrees Celsius. The report noted that, during the 2000s, China's own emissions were responsible for almost 60 percent of the global increase. It would "not be easy" for it to "drastically reduce" its emissions, but this was nonetheless considered to be both possible and necessary. The report recommended "bold and ambitious action" to ensure emissions peaked as early as possible and preferably by 2015. If China's emissions did not peak until 2025, the required emissions reduction would be "virtually unattainable".[85] The second report developed SEI's Deep Carbon Reduction Scenario in greater detail, examining the technical feasibility of "massively reducing" emissions by 2050 (to about 85 percent of the level in 1990). If the Scenario was achieved emissions would peak in 2017.[86]

The pre-Copenhagen period saw some manifestations of support within China for specific national emissions reduction targets. Hu Angang and Guan Qingyou asserted that China had a responsibility to play "a major role in the global fight against climate change". They advocated a "three-stage roadmap" in which China would aim for its carbon emissions to peak by 2020, or no later than 2030, to reduce to 1990 levels by 2030, and to reduce further to half of 1990 levels by 2050.[87] The target set by the government-supported Chinese Academy of Sciences was considerably less demanding, however, with emissions peaking between 2030 and 2040.[88]

The search for a date by which China might reach peak emissions was given greater credibility by a landmark new forecast by Chinese experts of its energy and emissions outlook. Although lacking official endorsement this report was of major significance. Titled the *2050 China Energy and CO_2 Emissions Report*, it totalled, in Chinese, almost 900 pages. It warned of a massive potential threat to China from climate change. It admitted that China's emissions had overtaken those of the US and proposed policies that could slow emissions growth by 2020, reach a peak around 2030, then possibly fall by 2050 to 2005 levels or lower. It proposed financial and price reforms, a carbon tax, and steps toward a cap-and-trade system. In the study, Jiang Kejun from ERI warned that under a business-as-usual approach carbon emissions could almost double by 2040, compared with an increase of only about one third under a *Low Carbon* scenario or even less under an *Enhanced Low Carbon* scenario. He described a low carbon strategy as "difficult but doable".[89]

This report was followed by a similar pre-Copenhagen forecast compiled by a task force that included many leading Chinese experts and international participants. It described a low carbon economy as "inevitable" for China. It did not recommend specific limits on absolute emissions but some of its scenarios included the "possibility" of reaching peak emissions. It presented four scenarios – *Business-as-Usual*, *High GDP Low Carbon*, *High GDP Enhanced Low Carbon*, and *Low GDP Low Carbon* – each detailing the technologies required, the economic implications, and the related emissions. China, the report stated, faced "a major strategic choice" – to "retake the road of developed countries" or to take a low carbon pathway that would reduce annual energy consumption by almost 20 percent and emissions by almost 30 percent. After 2030, if this pathway was adopted, China's absolute emissions would peak. The task force also recommended targets for 2050 to reduce energy intensity by 75 to 85 percent and carbon intensity by 85 to 90 percent.[90]

Following China's 2009 announcement of its new goal to reduce carbon intensity in 2020 by 40 to 45 percent compared to 2005, a new debate sprang up about whether this seemingly impressive target would be difficult or relatively easy for China to achieve. Many asserted it would indeed be challenging. Fatih Birol, the IEA's chief economist, stated that the new target meant that China had accepted responsibility for over 25 percent of the worldwide emissions reductions needed to limit warming to 2 degrees Celsius – more than its share on a population basis. The target was also in line with the IEA's 2010 scenario for stabilising atmospheric CO_2 at 450 ppm, suggesting China was "pulling its weight" toward achieving this global target.[91] LBNL was slightly more cautious. It judged that China's 2010 energy intensity goal was likely to be nearly achieved, and that it was "highly likely" that its new 2020 carbon intensity goal would attract "a similarly serious commitment and effort". It added, however, that use of a carbon *intensity* measure – rather than one related to *absolute* emissions – obscured the level of effort required and would allow emissions to continue to grow.[92] Others were even more sceptical. Michael Levi at the US Council on Foreign Relations saw China's target as "disappointing", "business-as-usual", "conservative", and showing "low apparent ambition". Several agencies, he said, had already projected cuts of 40 percent or more by 2020. The IEA had recommended a 48 percent cut and Levi agreed with McKinsey that a 54–56 percent reduction was achievable.[93]

Several experts came strongly to China's defence, however. William Chandler and Wang Yanjia thought China's new carbon intensity target was "severe, even draconian" and imposed efficiency standards "to a degree unparalleled in any other country". It had already forced the closure of "tens of thousands" of factories, power plants, and production lines. For China to bind itself to such a target was "surprising" and reflected its "fear" of climate change. Criticism, they asserted, could be explained only by "lazy scholarship" or "reflexive 'China bashing'". China's 40 to 45 percent target would require it to reduce its carbon intensity by 4 percent annually. The idea that this represented nothing more than business-as-usual was "demonstrably false".[94] David Stern and Frank Jotzo agreed that China's target would require ambitious mitigation policies. They estimated that a business-as-usual scenario would produce only a 24 percent reduction.[95] ZhongXiang Zhang also believed that meeting its target would be "challenging and costly" for China, and considered that, if it succeeded, it would contribute an impressive 31.6 percent of the global reductions required under the 450 ppm scenario.[96] David Cohen-Tanugi described China's target as "almost" in line with the level required, according to the IEA, to keep atmospheric GHGs below 450 ppm.[97] He considered three scenarios: a *Previous Domestic Commitments* scenario that would result in only a 37 percent carbon intensity reduction between 2005 and 2020; an *Extended Efforts* scenario leading to a 48 percent reduction; and an *Economic Restructuring* scenario that could reduce carbon intensity by up to 57 percent.[98]

The post-Copenhagen period saw the completion of several additional forecasts of China's prospects for curbing energy and carbon intensities. The joint UNDP/Renmin report cited earlier presented three low carbon scenarios based on a "Technological Optimization Model" developed by Renmin. In the *Business-as-Usual* scenario, emissions would increase rapidly and not peak until after 2050; in the *Emissions Control* scenario, new measures would be adopted, but emissions would still probably not peak before 2050; but the *Emissions Abatement* scenario would set a target to reach peak emissions by 2030, which was considered difficult but technologically possible. Notably, both the *Emissions Control* and *Emissions Abatement* scenarios envisaged carbon intensity dropping by more than China's official target.[99]

A 2010 World Bank report, with a whole-of-Asia coverage, indicated that under a *Sustainable Energy Development* scenario China's CO_2 emissions might peak by 2025. This assumed achievement of a "daunting" 4.3 percent annual decline in energy intensity. The report lauded the scale of China's achievements to date, describing its energy intensity reduction from 2005 to 2010 as "the most aggressive emissions reduction target in the world". It commented that China was installing new supercritical coal-fired power plants at only slightly greater cost than traditional plants, and already accounted for 75–80 percent of the world's installed capacity of supercritical and ultrasupercritical plants.[100]

China's own research institutes also continued to explore scenarios for accelerated development of a low carbon economy. A working paper by ERI's Jiang Kejun included a *Baseline* scenario reflecting existing policies and measures; a *Low Carbon* scenario with structural reform, dissemination of available technologies, "aggressive diversification" of the fuel mix, and "aggressive standards" for energy efficiency; and an *Enhanced Low Carbon* scenario that explored new technologies, maximised use of renewables and nuclear power, and introduced Carbon Capture and Storage. Under the two Low Carbon scenarios, CO_2 emissions would begin to plateau after 2020 and peak in 2030. Only under the Enhanced Low Carbon scenario would emissions fall sharply after 2030.[101]

A 2011 report by LBNL pursued an interesting new approach. It developed scenarios that, instead of concentrating on energy *supply*, focused on the determinants of future energy *demand*. Mark Levine saw insufficient attention in conventional forecasts to changing influences on energy demand as a "serious omission".[102] LBNL's new approach gave it the confidence to forecast that China's emissions would not grow indefinitely but would level off and decline, because "saturation" of demand would "peak around 2030". The report presented two main scenarios: a baseline or *Continued Improvement* scenario (CIS), and an enhanced energy efficiency or *Accelerated Improvement* scenario (AIS). Both indicated that energy consumption would approach a plateau around 2040, but CO_2 emissions would plateau or peak in 2025 under AIS and in 2030 under CIS.[103]

A fifth post-Copenhagen forecast was released in 2012 by the UK's Grantham Institute for Climate Change. Using a normative approach, it examined the existing and new energy technologies that could allow China to "significantly decarbonise" its economy by 2050. Noting that China's emissions had more than doubled since 2000, the report warned that with business-as-usual it could contribute nearly 30 percent of global emissions by 2050. It concluded, however, that there were "several feasible pathways" for emissions reductions that would be "broadly consistent" with limiting global warming to 2 degrees Celsius.[104]

The relative optimism of these post-Copenhagen reports is evident. But other commentators emphasised the huge environmental pressures that still confronted China, despite its evident good intentions. Nicholas Stern acknowledged that China was at the forefront of development of low carbon technologies and that a low carbon economy was a central priority of the Twelfth Plan. He estimated, however, that, even if China achieved its carbon intensity targets up to 2020, and made similar progress between 2020 and 2030, its emissions alone in 2030 (if GDP growth averaged 7 percent) would already be almost half of the maximum global emissions consistent with avoiding a temperature rise above 2 degrees Celsius. He recommended that China should therefore aim to cut its emissions intensity by 50 percent each decade or 29 percent in each five year plan. Stern commented that this would require "a stronger level of ambition than currently envisaged" in China's targets.[105]

On an equally sombre note, Germany's Hanns Maull pointed out that the IEA's 2010 *World Energy Outlook* estimated – even in its optimistic *New Policies* scenario – that China could account for about 35 percent of the net increase in world energy demand to 2035, and it could also account for 58 percent of the growth in global CO_2 emissions in that period. Its electricity demand was projected to triple between 2008 and 2035. The IEA had described this expansion as "staggering". By 2035 China could account for 28 percent of global emissions, compared to 12.5 percent for the US and 7.8 percent for the EU. However, Maull accepted that China would continue to do "a great deal" to help stem global warming and that its emissions might peak before mid-century.[106] The IEA's 2011 *World Energy Outlook* also offered little comfort, noting that, from consuming half as much energy as the US in 2000, China had become the world's biggest consumer, and by 2035 it was projected to consume almost 70 percent more than the US.[107]

Given these indications that even China's best efforts to transition to a low carbon economy might indeed be overwhelmed by continued rapid emissions growth, it is not surprising that it remained concerned about its environmental prospects. In January 2012 its scientific advisors, in a *Second National Assessment Report on Climate Change*, warned that China faced "extremely grim" conditions as a consequence of global warming. Without counter-measures, its grain output could fall by 5 to 20 percent by 2050. By 2100, its average temperatures could rise, under various scenarios, by between 2.5 and 4.6 degrees. Many of its provinces and cities could face severe or chronic water shortages, the retreat of its glaciers could accelerate, and rising sea levels could threaten some of its biggest cities and export zones. Even so, the report stated, China's CO_2 emissions would only begin to fall after about 2030, with big falls only after 2050.[108]

Evaluating environmental forecasts

The quantitative forecasts reviewed above represent, in aggregate, a major national and international effort by statistical modellers to provide guidance to decision-makers about China's potential future emissions pathways, giving them the opportunity to plan and act appropriately. These separate and independent forecasts were especially valuable because they approached the task from several different perspectives. Even Spyros Makridakis and Nassim Taleb conceded that averaging the forecasts of several models tends to improve accuracy and reduce variance in forecasting errors.[109] The reports also conformed to Vaclav Smil's recommendations that specific single-point forecasts be avoided in favour of developing alternative scenarios, and that forecasters should include normative scenarios that would indicate the policy measures and technologies required to meet pre-set targets.[110]

It was also encouraging that independent efforts were made to compare the methods and findings of these different forecasts. Perhaps the most detailed comparison was conducted by Li Huimin and Qi Ye at Tsinghua University. They compared the reports discussed above by McKinsey, Tyndall, ERI, Renmin/UNDP, LBNL and the IEA, highlighting significant and "confusing" variations due in large part to their different methods and assumptions. They found, for example, that emissions in 2050 for the scenario with the highest emissions were 3.7 times those for the scenario with the lowest emissions. Projections of the timescale for reaching a peak in emissions also varied greatly, with most falling within the 2030 to 2050 range or later. However, they made the very important observation that most scenarios based on accelerated mitigation efforts were in agreement that these were likely to lead to much slower emissions growth and a steady decline after about 2030. Li and Qi concluded that although the results of these studies were "very uncertain", nevertheless they all indicated the feasibility of continued reductions in emissions intensity.[111]

Similar comparisons were made by LBNL and the Grantham Institute. LBNL noted that the forecasts all suggested China's energy and CO_2 emissions were unlikely to peak before 2030, although growth was expected to slow after 2020 in the abatement scenarios. They also assumed that China would meet its 40 to 45 percent emissions reduction target by 2020. All the forecasts highlighted the clear benefits from continued aggressive decarbonisation of China's economy. LBNL commented that each new modelling approach added new insights, and that "robust and credible" models could play a key role in informing policy-makers. It expressed concern that even "slightly different" modelling assumptions could result in "significant discrepancies", but despite such differences NBNL found it significant that five of the major modelling studies showed a "general clustering" or agreement in their projections of China's future energy consumption and emissions.[112]

Considering all the scenarios together, however, they revealed large variations in forecasted emissions outcomes, especially to 2050. The scenarios can be grouped approximately into four categories – first, baseline scenarios that assumed continued rapid growth in emissions; second, abatement scenarios that assumed vigorous but still relatively modest mitigation efforts; third, enhanced abatement scenarios that assumed timely, determined and ambitious efforts; and fourth, deep reduction scenarios that would require immediate and aggressive reduction efforts and be very challenging economically, politically, and technologically. Most of the modelling teams (including IEA, Tyndall, McKinsey, SEI and Grantham) argued that only the fourth group of scenarios would give any confidence that China could meet global climate mitigation objectives. This fourth group would require China's carbon emissions to peak by 2020, while the third group saw emissions peaking around 2030, and the second group saw emissions peaking before 2050.

Collectively the reports provided the best publicly available quantitative estimates of China's environmental trends. Significantly, many of them were somewhat more optimistic in many respects than might perhaps have been expected. They highlighted the potential for China to take action, supported internationally, to mitigate the worst of the possible consequences of its growth for global climate change. Such attempts to forecast China's environmental future quantitatively can be viewed in two ways. On the one hand, these projects are laudable for their ambitious scope and apparent precision. They potentially offer unique insights – not possible through any other approach – into vitally important questions such as: what impact specific measures might have on managing future energy demand and carbon emissions; when China might be able to reach a peak in its annual emissions and then reduce them; and whether China will be able to make a sufficient contribution to global efforts to manage climate change. On the other hand, such modelling has in the past proved to be notoriously unreliable, as shown only too clearly by its abject failure to anticipate the paradigm-changing effects of the surge in China's energy demand in the early 2000s. It is little wonder that the findings of such forecasts are still considered speculative and that their reliability is debated.

Within the forecasting profession, Charles Doran observed that forecasts ultimately fail because "no technique has been developed that allows the forecaster to predict … when a nonlinearity will occur".[113] Vaclav Smil pointed out that energy and environmental forecasting offered some of the best examples available of failed predictions – not just because the forecast values were "significantly off", but because the "uselessness" of such forecasts aiming to look from five to 50 years ahead often became obvious within just a few years or even months.[114] He believed that the long chains of assumptions used in increasingly complex models, as part of their quest for realism, had become self-defeating. Forecasts would continue to fail because many critical variables could not be anticipated or estimated. In circumstances of high uncertainty, he quipped that complex models were unlikely to produce better results than a small electronic calculator or jottings on the back of an envelope.[115]

It remains to be seen whether the degree of optimism displayed in many recent forecasts of China's environmental trends will be regarded with hindsight as unrealistic or misleading. Makridakis and Taleb criticised academics for encouraging expectations among decision-makers that sophisticated models could provide accurate assessments of future uncertainties. Statistical models, they argued, sometimes underestimated uncertainty catastrophically. In complex social systems, events were also interdependent and were affected greatly by human actions and reactions.[116] Makridakis, noting that the "one certainty" was that forecasts are "usually wrong", advised modellers to deal with this situation by following three steps: "(1) 'Accept' the existence of uncertainty; (2) 'Assess' its magnitude; and (3) 'Augment' its range". As a simple rule, he suggested that forecasters should estimate the upper and lower range of their predicted values and then double that range.[117]

Fortunately modellers of China's environmental trends have not adopted an excessively cautious or defeatist attitude. They have accepted a responsibility to provide the best advice they can. By refusing to be defeated by uncertainty, it seems clear that they have succeeded in providing invaluable assistance to China's decision-makers. It is difficult, for example, to imagine how China would have been able to announce its specific and ambitious commitments to reduce energy and carbon intensity without the information and reassurance provided by forecasters and their models. Nor would it seem likely without such modelling results that China would have begun to talk, despite all the evident constraints, about possible timeframes within which it might reach a peak in emissions. Some of its advisers have recommended advancing the timetable for peak emissions. Jiang Kejun, for example, pointed to the more rapid than expected development of renewable energy as reason to expect that emissions may peak by 2025.[118] Other reports suggest that China may have also begun to consider interim targets for absolute emissions limits.[119]

Makridakis's assessment that modellers routinely underestimate the extent of uncertainty in long-term forecasting suggests nevertheless that forecasters should be more imaginative and exploratory when considering specific uncertainties, both negative and positive, that may impact on their findings. In China, it is possible to speculate about several specific negative uncertainties that could limit the effectiveness of its efforts to reduce emissions. These include, for example, the likelihood that necessary changes will continue to be resisted by vested interests in business and government that give greater priority to economic growth, profitability, and personal enrichment. There could also be public resistance to measures that increase the cost of living, and a loss of momentum due to perceptions that the easy gains in emissions reduction – the "low-hanging fruit" – have already been harvested. Unforeseen crises – such as economic slowdown or recession, political unrest, and international conflict – could also undermine the achievability of ambitious scenarios.

Most concerning among all of these negative uncertainties is the possibility that China's continued strong GDP growth may offset the effectiveness of even the most vigorous measures to contain energy demand and emissions growth, which of course is at the crux of the second critical uncertainty considered in this chapter. In 2011, for example, China's GDP grew by 9.2 percent and the IEA estimated its emissions increased by 9.3 percent, spurring a 3.2 percent rise in global emissions despite falls in the US and Europe.[120] Consequently China was unable to achieve its energy intensity target in 2011.[121] On the other hand, *slower* GDP growth could also have negative consequences, especially by reducing the resources available for investment in low carbon development. Another potential source of large forecasting errors may be statistical underestimation of China's actual carbon emissions. Researchers published an article in *Nature* magazine that drew attention to differences in China's national and provincial level carbon emissions data. They concluded that reporting of China's emissions in 2010 had been underestimated by 1.4 gigatonnes – equivalent to Japan's total annual CO_2 emissions.[122]

Forecasters also need, however, to take into account uncertainties that have potentially large *positive* impacts. In this context it is notable that there are several very powerful motivations, if not imperatives, for China to pursue a low carbon and environmentally sustainable future. First, its energy security will be threatened if it does not preserve domestic fossil-fuel energy reserves and reduce its over-dependence on imported energy sources. Second, it is highly motivated to improve the health and quality of life of its citizens by reducing atmospheric and other forms of pollution – especially after the media attention given to record levels of air pollution in Beijing.[123] Third, it seeks opportunities to restructure its economy, including by capitalising on the potentially vast commercial gains to be made from accelerated development of a low carbon economy. Last but not least, China acknowledges the need to protect its own environmental interests and the future of the planet by contributing to global measures to fight climate change. China's decision-makers and modellers appear to be well aware of all of these imperatives. Its leadership has shown considerable political resolve, including a willingness to impose challenging new environmental targets and actions, and a capacity for determined implementation of them. While it is likely that China's still rapidly growing emissions will remain the major single source of global emissions growth, its innovative environmental leadership also suggests it will be a major contributor to global solutions.

Cleaner fossil-fuel sources and technologies may also assist to contain emissions, perhaps to an extent not previously anticipated. Experience in the United States, for example, suggests the potentially dramatic potential for exploiting unconventional gas deposits. The huge expansion of shale gas extraction in the US has, according to the IEA, led to its energy-related carbon emissions falling over a five year period by 450 million tonnes (compared to a 700 million tonnes increase in China in 2011 alone).[124] The Twelfth Five Year Plan reportedly aims for shale gas to provide 10 percent of China's energy by 2020.[125] Expanded non-fossil fuel power generation also has very promising potential in China, including especially the planned huge and rapid expansion of cheaper and safer nuclear power. In addition, China evidently has great hopes for continued exponential growth of wind power, including offshore wind. A joint ERI and IEA report has estimated that wind's share in China's electricity production could rise from 1 percent today to 17 percent by 2050.[126] China's massive investment in wind and solar systems is relentlessly driving down unit costs, making the use of these renewable technologies much more attractive domestically and globally. Last but not least China's introduction of pilot carbon emissions trading schemes, followed perhaps by a national scheme, may also have a huge impact.

These potential developments, which could affect China's environmental trends both negatively and positively, underline the major uncertainties that continue to confront efforts by modellers to forecast future trends. But such uncertainties do not seem likely to fundamentally undermine the value of those forecasts to decision-makers. Uncertainty is an inherent part of the forecasting process, but that process is iterative. Forecasters can attempt to learn from previous mistaken assumptions and be alert for new trends and future surprises. It is also difficult to dispute Vaclav Smil's view that it will not be possible in any case "to prevent our forecast-happy, computer-equipped civilization from issuing a ceaseless stream of outlooks and predictions and scenarios".[127]

Environmental futures

The exploration above of two major critical uncertainties in China's environmental future reveals that expert opinions have evolved significantly. It may be useful here to summarise these developments. During the 1980s and 1990s, the prevailing consensus gave rise to serious concerns about the steady deterioration of the environment. Experts were very sceptical about the Chinese government's willingness to address this issue with sufficient vigour. Concern quickly turned to alarm following the revelation that energy intensity had soared in the early 2000s. This raised the further gloomy prospect that China's emissions would undermine international efforts to counter global warming.

Most China watchers were very slow, however, to become aware of the impressive scale of the government's reaction. Some experts continued to be very pessimistic about China's environmental performance, while others who followed its actions more closely saw cause for hope. Many of the latter were inclined to defend China's progress despite its declining image in international climate change negotiations. China's domestic targets to reduce energy and carbon intensity were considered likely to have an important beneficial impact. Others drew attention to its unexpectedly rapid progress in developing a low carbon economy and in becoming a world investment leader in this field. China's reputation for environmental management was still mixed, but it attracted many supporters as well as critics among environmental experts.

At the same time China's carbon emissions have nonetheless continued to grow strongly, making it the world's largest emitter and the world's biggest contributor to the growth of global emissions, and threatening to undo its creditable progress in reducing emissions intensity. In response, from the mid-2000s, a number of both Chinese and international forecasters turned their attention to modelling its future emissions trends and calculating the scope for progress in reducing emissions under various scenarios. These scenarios ranged from (unlikely) business-as-usual scenarios to (very challenging) normative scenarios that would make an appropriate contribution to international targets for containing global warming. Such forecasts (including those by China's own research agencies) indicated a surprisingly high level of ambition and even of cautious and qualified optimism among forecasters about the feasibility of achieving an early plateau or peak in China's absolute emissions, followed by steady reductions over the longer term. Those forecasts appeared to increase the Chinese government's confidence, and determination, to maintain progress in controlling emissions. Much uncertainty remains, however, about whether such efforts will be enough to counter still-rising emissions and make a sufficient contribution to reducing global atmospheric GHG concentrations.

Despite their ominous and potentially catastrophic implications, it is puzzling that these critical uncertainties in the environmental domain have tended to be comparatively less studied and less frequently discussed among mainstream China experts than uncertainties in the other three domains. There is only a low level of awareness of the issues being debated among specialist environmental forecasters or of the shifting trends in their views. Partly this may be because expert studies in the environmental domain are somewhat less publicly accessible, partly because the discussion is complex and technical, and partly also because the trends themselves are changing rapidly making it difficult to keep up-to-date. This is very unfortunate because a plausible case can be mounted that uncertainties in China's environmental domain are more critical than in any other, especially if one accepts the prevailing judgement of the great majority of scientists, governments and international organisations that early action to curb GHG emissions is vital for the future viability of the world's climate, ecology and economy.

The consequence of inadequate public awareness of environmental developments and uncertainties in China is that perceptions are often inconsistent with the changing realities. There is awareness that China's environmental problems are huge, and that it is the chief source of the current growth in global emissions. There appears also, however, to be a common and largely mistaken perception that the Chinese government's response – and its participation in international climate change negotiations – is woefully inadequate. There seems to be an accompanying widespread feeling that there is little point in other countries taking vigorous measures to contain GHG emissions if China is not willing to act with equal vigour. In contrast with these popular perceptions, the expert views surveyed in this chapter suggest that China – despite being clearly the major current contributor to the climate change crisis – is acting with considerable vigour and determination, and shows some prospect of beginning to curb the absolute level of its emissions. It seems indisputable, therefore, that it is in the interests of all countries for China's efforts to receive greater recognition, encouragement and support. It is also apparent that there is a very strong case for the critical uncertainties presented in this chapter to be kept under continuous and intensive scrutiny. Progress is evident, but success is by no means assured.

Notes

1 Vaclav Smil, *The Bad Earth: Environmental Degradation in China* (New York: M. E. Sharpe, 1984), xii, 168, and 198–200. See also Vaclav Smil, "China's Environmental Morass", *Current History* (September 1989), 277–288.
2 Vaclav Smil, *China's Environmental Crisis: An Inquiry into the Limits of National Development* (New York: M. E. Sharpe, 1993), 188–203.
3 This proved to be an underestimation when China became the largest emitter in 2006 or 2007.
4 Vaclav Smil, *China's Past, China's Future: Energy, Food and Environment* (New York: Routledge Curzon, 2004), 24–26, 120–121, 152, 178, and 202–203.
5 He Bochuan, *China on the Edge: The Crisis of Ecology and Development* (San Francisco, CA: China Books and Periodicals, 1991), xiv-xv and 193.
6 Elizabeth C. Economy, *The River Runs Black: the Environmental Challenge to China's Future* (Ithaca, NY: Cornell University Press, 2005), 1–26 and 257–274.
7 *Der Spiegel* (7 March 2005).
8 Peter Aldhous, "China's Burning Ambition", *Nature* 435 (30 June 2005), 1152–1154.
9 Jianguo Liu and Jared Diamond, "China's Environment in a Globalizing World", *Nature* 435 (30 June 2005), 1179–1186.
10 Nathan Nankivell, "The National Security Implications of China's Emerging Water Crisis", *China Brief* 5, no. 17 (2 August 2005), and Zmarak Shalizi, *Addressing China's Growing Water Shortages and Associated Social and Environmental Consequences* (World Bank, April 2006).
11 Peter Engardio, *Chindia: How China and India are Revolutionising Global Business* (New York: McGraw Hill, 2006), 297, 305 and 310.
12 Peter Ho and Eduard B. Vermeer, eds. *China's Limits to Growth: Greening State and Society* (Malden, MA: Wiley-Blackwell, 2006), vii, 7 and 15–21. Ho discussed the problem of "greenwash" where a thin "green veneer" was used to disguise environmentally unfriendly development.
13 Campbell and Darsie, *China's March on the 21st Century*, 53–60.
14 Daniel H. Rosen and Trevor Houser, "What Drives China's Demand for Energy (and What it Means for the Rest of Us)", in *The China Balance Sheet in 2007 and Beyond*, ed. C. Fred Bergsten *et al.*, 25–31 and 45–51; Daniel H. Rosen and Trevor Houser, "Energy Implications of China's Growth", in *China's Rise: Challenges and Opportunities*, ed. C. Fred Bergsten, *et al.*, 140–162. See also Daniel H. Rosen and Trevor Houser, *China Energy: A Guide for the Perplexed* (China Balance Sheet, May 2007).
15 Noureddine Berrah *et al.*, *Sustainable Energy in China: The Closing Window of Opportunity* (Washington, DC: World Bank, March 2007), xiii–xx, xxxii–ix, 30–32 and 71.
16 Elizabeth Economy, "China vs. Earth", *The Nation* (7 May 2007).
17 Elizabeth C. Economy, "The Great Leap Backward?", *Foreign Affairs* 86, no. 5 (September/October 2007), 38–59. See also: Elizabeth Economy and Kenneth Lieberthal, "Scorched Earth: Will Environmental Risks in China Overwhelm its Opportunities?", *Harvard Business Review* (June 2007), 88–96.
18 Malcolm Shealy and James P. Dorian, *Growing Chinese Energy Demand: Is the World in Denial?* (Center for Strategic and International Studies, October 2007).
19 William Chandler, *Breaking the Suicide Pact: US–China Cooperation on Climate Change* (Carnegie Endowment for International Peace, March 2008). Other passionately argued proposals for cooperation between China and the United States to address climate change included Kenneth Lieberthal and David Sandalow, *Overcoming Obstacles to US–China Cooperation on Climate Change* (Washington, DC: John L. Thornton China Center at Brookings, January 2009), and *Common*

Challenge, Collaborative Response: A Roadmap for US–China Cooperation on Energy and Climate Change (Pew Center on Global Climate Change & Asia Society Center on US–China Relations, January 2009).

20 Kelly Sims Gallagher, "China Needs Help with Climate Change", *Current History* (November 2007), 389–394.

21 Kelly Sims Gallagher, "The Voracious Dragon: Environmental Implications of China's Rising Energy Consumption", in *Global Giant: Is China Changing the Rules of the Game?*, ed. Eva Paus, Penelope B. Prime, and Jon Western (New York: Palgrave Macmillan, 2009), 93–109.

22 Paul G. Harris, "China's Road to Destruction: Following the West on Global Warming", *Global Asia* 3, no. 4 (Winter 2008), 88–94.

23 Jiang Lin *et al.*, *Achieving China's Target for Energy Intensity Reduction in 2010: An Exploration of Recent Trends and Possible Future Scenarios* (Ernest Orlando Lawrence Berkeley National Laboratory, December 2006). This report was reissued in the following year with the more upbeat title of: *Taking Out One Billion Tons of Carbon: the Magic of China's 11th Five Year Plan.*

24 Mark D. Levine and Nathaniel T. Aden, *Global Carbon Emissions in the Coming Decades: The Case of China* (Lawrence Berkeley National Laboratory, 2008), 28–29.

25 Mark D. Levine, *Testimony Before the US–China Economic Review Commission Hearing on China's Energy Policies and Their Environmental Impacts* (13 August 2008).

26 This was followed by issue of *China's National Climate Change Programme* (2007), *China's Scientific and Technological Actions on Climate Change* (2007), and *China's Policies and Actions for Addressing Climate Change* (2008). Such documents were regularly updated in later years. See also: Roger Irvine, *Victim of Success: China's Growth and Environmental Consequences* (Sydney: Lowy Institute for International Policy, December 2009), 6, and *Climate Change Mitigation Measures in the People's Republic of China* (Pew Center on Global Climate Change, April 2007).

27 Joanna I. Lewis, "China's Climate Change Strategy", *China Brief* 7, no. 13 (27 June 2007).

28 ZhongXiang Zhang, "China's Reds Embrace Green", *Far Eastern Economic Review* (June 2007).

29 Ross Garnaut, Frank Jotzo, and Stephen Howes, "China's Rapid Emissions Growth and Global Climate Change Policy", in *China's Dilemma: Economic Growth, the Environment and Climate Change*, ed. Ligang Song and Wing Thye Woo (Australian National University E Press, 2008), 184–187.

30 Karl Hallding, Guoyi Han, and Marie Olsson. "China's Climate– and Energy–Security Dilemma: Shaping a New Path of Economic Growth". *Journal of Current Chinese Affairs* 38, no. 3 (2009), 119–134. See also Karl Hallding and Marie Olsson, *Balancing Climate Concerns and Energy Security: China searching for a new developmental pathway* (Stockholm Environment Institute, 2010).

31 Bram Buijs, *China, Copenhagen and Beyond: The Global Necessity of a Sustainable Energy Future for China* (Netherlands Institute of International Relations, 2009), 9, and Michael T. Klare, Penelope B. Prime, and Jon Western, "Fueling the Dragon: China's Strategic Energy Dilemma", in *Global Giant: Is China Changing the Rules of the Game?*, ed. Eva Paus (New York: Palgrave Macmillan, 2009), 197.

32 Jianjun Tu, "Future Prospects of China's Policy on Climate Change", *China Brief* 9, no. 1 (12 January 2009).

33 Xie Zhenhua, *US–China Climate Change Cooperation*, remarks at Carnegie Endowment for International Peace (Washington DC, 18 March 2009).

34 Todd Stern, *China and the Global Climate Challenge* (Center for American Progress, 3 June 2009).

35 ZhongXiang Zhang, "Is it fair to treat China as a Christmas tree to hang everybody's complaints? Putting its own energy saving into perspective", *Energy Economics* 32 (2010), S48-S56. See also ZhongXiang Zhang, *Climate Commitments to 2050: A Roadmap for China* (Honolulu, HI: East–West Center, 4 December 2009).
36 *China's Clean Revolution* (The Climate Group, August 2008).
37 *China's Clean Revolution II: Opportunities for a Low Carbon Future* (The Climate Group, August 2009).
38 *The China Greentech Report 2009* (The Greentech Initiative and MangoStrategy, September 2009), 15 and 137. See also ZhongXiang Zhang, "China in the Transition to a Low-Carbon Economy", *Energy Policy* 38 (2010), 6638–6653.
39 *Cleaner Coal in China* (International Energy Agency, 20 April 2009), 22 and 26.
40 Keith Bradsher, "China Outpaces US in Cleaner Coal-fired Plants", *New York Times* (11 May 2009).
41 Julian Wong and Andrew Light, *China Begins Its Transition to a Clean-Energy Economy: China's Climate Change Progress by the Numbers* (Center for American Progress, 4 June 2009).
42 Deborah Seligsohn, "China and cclimate change Doing more than you think", *China Economic Quarterly* (September 2008).
43 Deborah Seligsohn, Fei Feng, Roland Priddle and Leiping Wang, *China, the United States, and the Climate Change Challenge* (Washington, DC: World Resources Institute, October 2009).
44 Hu Angang, "Green Development: the inevitable choice for China", *China Dialogue* (26 June 2006).
45 Hu Angang, "A new approach at Copenhagen", *China Dialogue* (6 April 2009).
46 Development Research Center of the State Council, "Greenhouse gas emissions reduction: A theoretical framework and global solution", in *China's New Place in a World in Crisis: Economic, Geopolitical and Environmental Dimensions*, ed. Ross Garnaut, Ligang Song, and Wing Thye Woo (Australian National University E Press, 2009).
47 Jonathan Watts, "China considers setting targets for carbon emissions", *Guardian* (19 April 2009). See also, *China Sustainable Development Strategy Report 2009: China's Approach towards a Low Carbon Future* (Chinese Academy of Sciences, 2009).
48 See for example Mark Lynas, "How do I know China wrecked the Copenhagen deal? I was in the room", *Guardian* (22 December 2009), and George Monbiot, "If you want to know who's to blame for Copenhagen, look to the US Senate", *Guardian* (21 December 2009).
49 Zhang, "China in the Transition to a Low-Carbon Economy", 110–113. See also Bjorn Conrad, "China in Copenhagen: Reconciling the 'Beijing Climate Revolution' and the 'Copenhagen Climate Obstinancy'", *China Quarterly* 210 (June 2012), 445.
50 Joanna Lewis, "Environmental Challenges: From the Local to the Global", in *China Today, China Tomorrow: Domestic Politics, Economy, and Society*, ed. Joseph Fewsmith (Lanham, MD: Rowman & Littlefield, 2010), 264–275.
51 *China and a Sustainable Future: Towards a Low Carbon Economy and Society* (China Human Development Report, Renmin University and UNDP, 2010), 75–96.
52 David Held, Eva-Maria Nag, and Charles Roger, *The Governance of Climate Change in China* (Preliminary Report, LSE-AFD Climate Governance Programme, January 2011), 49–53.
53 *Who's Winning the Clean Energy Race?: Growth, Competition and Opportunity in the World's Leading Economies* (G-20 Clean Energy Factbook, Pew Charitable Trusts, 2010).
54 Daniel Roth, "The Key to Fixing Global Warming?" *China Wired*, May 2010.
55 *The Green Economy: The Race is On*, (DB Climate Change Advisors, Deutsche Bank Group, 22 April 2010).

56 *Investing in Climate Change 2011*: *The Mega-Trend Continues – Exploring Risk and Return* (DB Climate Change Advisors, Deutsche Bank Group, February 2011).
57 *Delivering Low Carbon Growth: A Guide to China's 12th Five Year Plan* (The Climate Group, March 2011), 1–5 and 30.
58 *The China Greentech Report 2011: China's Emergence as a Global Greentech Market Leader* (The China Greentech Initiative, 2011), and *The China Greentech Report 2012: Faced with Challenges, China Accelerates Greentech Growth* (The China Greentech Initiative, 2012).
59 John Lee, "The greening of China a mirage", *The Australian* (19 September 2011).
60 John Lee, "China's Fake Carbon Tax", *The Wall Street Journal* (10 January 2012).
61 Alvin Lin and Yang Fuqiang, "China's carbon tax is very real", *China Dialogue* (27 January 2012).
62 Martin Adams, ed. *A greener shade of grey: A special report on renewable energy in China* (Hong Kong: Economist Intelligence Unit, 2012), 2–7.
63 Todd Johnson, *Clear Water, Blue Skies: China's Environment in the New Century* (Washington, DC: World Bank, 1997), 1–3 and 29–40. See also Kristof, "The Rise of China", 64–65, and Wen-Yuan Niu and William M. Harris, "China: the Forecast of its Environmental Situation in the 21st Century", *Journal of Environmental Management* 47 (1996), 101–112.
64 Robert Priddle, "China's Long-term Energy Outlook", in *China in the 21st Century*: *Long-term Global Implications* (OECD, 1996), 95–99 and 121–125. The IEA included several caveats, noting that its forecasting model was based on many assumptions and on scarce data, and that standard techniques of extrapolation were "highly dubious" given the unknown affects of future policy measures and the unpredictable behaviour of economic agents.
65 *Cooperation in the Energy Futures of China and the United States* (Washington, DC: National Academies Press, 2000), 15–20 and 47–48. The study claimed, rather dubiously, that "significant progress" had been made in forecasting energy demand using increasingly complex modelling systems that achieved "greater accuracy of predictions". See also, *Energy Futures and Urban Air Pollution: Challenges for China and the United States*, ed. National Academy of Sciences (The National Academies, 2007).
66 Zhou Dadi and Mark Levine, *China's Sustainable Energy Future: Scenarios of Energy and Carbon Emissions* (Lawrence Berkeley National Laboratory, 2004), 1–4 and 13–27. Subsequent estimates fell within a similar range.
67 William Chandler and Holly Gwin, *China's Energy and Emissions: A Turning Point?* (Pacific Northwest National Laboratory, 2004). Another report by Japan's Institute for Energy Economics also expressed concern. It presented a reference scenario which suggested that both energy consumption and CO_2 emissions in China could approximately triple by 2030: see Zhidong Li, Kokichi Ito, and Ryoichi Komiyama, *Energy Demand and Supply Outlook in China for 2030* (Institute of Energy Economics Japan, 2005).
68 Rosen and Houser, "What Drives China's Demand for Energy", 25–31 and 45–51; Rosen and Houser, "Energy Implications of China's Growth", 140–162. See also Rosen and Houser, *China Energy*: *A Guide for the Perplexed*, and Frank Wang and Hongfei Li, "Environmental Implications of China's Energy Demands: An Overview", in *China's Environment and the Challenge of Sustainable Development*, ed. Kristen A. Day (Armonk, NY: M. E. Sharpe, 2005), 180–195.
69 Warwick J. McKibbin, *Environmental Consequences of Rising Energy Use in China* (Sydney: Lowy Institute for International Policy, 2005).
70 Sandra Lawson, David Heacock, and Anna Stupnytska, *Why the BRICs Dream Won't Be Green* (Goldman Sachs, 18 October 2006), and Sandra Lawson, David Heacock, and Anna Stupnytska, *Why the BRICs Dream Should Be Green* (Goldman Sachs, 13 February 2007). See also Jiang Kejun, Xiulian Hu, Xianli Zhu, Amit Garg,

Kirsten Halsnaes and Qiang Liu, *Balancing Development, Energy and Climate Priorities in China: Current Status and the Way Ahead* (UNEP Riso Centre on Energy Climate and Sustainable Development, Roskilde, Denmark, September 2007).

71 Pan Jiahua, Chen Ying, Zhuang Guiyang and Zhao Xingshu, *Understanding China's Energy Policy: Economic Growth and Energy Use, Fuel Diversity, Energy/Carbon Intensity, and International Cooperation* (Beijing: Chinese Academy of Social Sciences, August 2006), 5–9 and 60–63. See also Lu Xuedu, Pan Jiahua, and Ying Chen, "Sustaining Economic Growth in China under Energy and Climate Security Constraints", *China and World Economy* 14, no. 6 (2006), 85–97.

72 Levine and Aden, *Global Carbon Emissions in the Coming Decades: The Case of China.*

73 Ibid.

74 See also the low levels of carbon emissions forecast in Detlef van Vuuren and Z. Fengqi, "Energy and emission scenarios for China in the 21st century – exploration of baseline development and mitigation options", *Energy Policy* 31 (2003), 369 and 372.

75 Peter Sheehan and Fiona Sun, *Energy Use and CO_2 Emissions in China: Retrospect and Prospect* (Centre for Stategic Economic Studies, Victoria University, Australia, June 2006). Notably, they qualified their forecast with the comment that the failure of earlier projections indicated the limited ability of advanced modelling techniques to predict future developments.

76 Peter Sheehan and Fiona Sun, *Energy Use in China: Interpreting Changing Trends and Future Directions* (Centre for Strategic Economic Studies, Victoria University, Australia, 2007).

77 Maximilian Auffhammer and Richard T. Carson, *Forecasting the Path of China's CO_2 Emissions: Offsetting Kyoto – and Then Some* (University of California, 2006). See also John Hawksworth, *The World in 2050: Implications of Global Growth for Carbon Emissions and Climate Change Policy* (PricewaterhouseCoopers, 2006), 44, and John Hawksworth, *The World in 2050: Can Rapid Growth be Reconciled with Moving to a Low Carbon Economy?* (PricewaterhouseCoopers, July 2008).

78 Zmarak Shalizi, "Energy and Emissions: Local and Global Effects of the Giants' Rise", in *Dancing with Giants: China, India and the Global Economy*, ed. L. Alan Winters and Shahid Yusuf (World Bank, 2007).

79 Warwick J. McKibbin, "Growth, energy use and greenhouse gas emissions in China," in *The Turning Point in China's Economic Development*, ed. Ross Garnaut and Ligang Song (Australian National University, 2006), 335–359.

80 *World Energy Outlook 2007: China and India Insights*, (International Energy Agency, 2007), 3. See also Rosen and Houser, "What Drives China's Demand for Energy".

81 Ning Zeng, Yihui Ding, Jiahua Pan, Huijun Wang and Jay Gregg, "Climate Change – the Chinese Challenge", *Science* 319 (8 February 2008). See also Zhuang Guiyang, "How Will China Move Towards Becoming a Low Carbon Economy?", *China and World Economy* 16, no. 3 (2008), 103.

82 Geoffrey J. Blanford, Richard G. Richels, and Thomas F. Rutherford, *Revised Emissions Growth Projections for China: Why Post-Kyoto Climate Policy Must Look East* (Harvard Kennedy School, September 2008).

83 Tao Wang and Jim Watson, *China's Energy Transition: Pathways for Low Carbon Development* (University of Sussex and Tyndall Centre for Climate Change Research, 2009), 2–5. See also Tao Wang and Jim Watson, "Scenario analysis of China's emissions pathways in the 21st centrury for low carbon transition", *Energy Policy* 38 (2010).

84 *China's Green Revolution: Prioritizing technologies to achieve energy and environmental sustainability* (McKinsey & Company, February 2009).

85 *Going Clean: The Economics of China's Low-carbon Development*, (Stockholm Environment Institute and Chinese Economists 50 Forum, November 2009), v–vi,

1–5 and 39. See also, *China's Pathway Towards a Low Carbon Economy*, (China Council for International Cooperation on Environment and Development, November 2009).

86 Charles Heaps, *A Deep Carbon Reduction Scenario for China* (Stockholm Environment Institute, 24 May 2009).

87 Hu Angang and Guan Qingyou, "Fighting Climate Change: China's Contribution", *China Dialogue* (8 and 10 October 2008), and Hu Angang, "A new approach at Copenhagen".

88 Jonathan Watts, "China considers setting targets for carbon emissions", *Guardian* (19 April 2009); *China Sustainable Development Strategy Report 2009: China's Approach towards a Low Carbon Future*; and Catherine Brahic, "China warns of huge rise in emissions", *New Scientist* (22 October 2008).

89 Reports by Chris Buckley, *Reuters* (17 August 2009).

90 *China's Pathway Towards a Low Carbon Economy*, 1–5, 70–78, 94–97 and 120.

91 Marlowe Hood, "China's climate pledge to meet a quarter of global needs: IEA", *AFP* (26 November 2009). See also *Delivering Low Carbon Growth: A Guide to China's 12th Five Year Plan.*

92 "Statement of the China Energy Group, Lawrence Berkeley National Laboratory on China's Recently Announced Carbon Intensity Target", undated. For similar comments see Bo Xu, Qie Sun, Ronald Wennersten and Nils Brandt, *An Analysis of Chinese Carbon Dioxide Mitigation Strategy* (Royal Institute of Technology, Sweden, June 2010).

93 Michael A. Levi, *Assessing China's Carbon-Cutting Proposal* (Council on Foreign Relations, 30 November 2009). Similar comments were made in Carlo Carraro and Massimo Tavoni, "Looking ahead from Copenhagen: how challenging is the Chinese carbon intensity target?", *Climate Science and Policy* (December 2009).

94 William Chandler and Yanjia Wang, *Memo to Copenhagen: Commentary is Misinformed – China's Commitment is Significant* (Carnegie Endowment for International Peace, December 2009).

95 David I. Stern and Frank Jotzo, "How ambitious are China and India's emissions intensity targets?", *Energy Policy* 38 (2010), 6776–6783.

96 ZhongXiang Zhang, "Assessing China's Energy Conservation and Carbon Intensity: How Will the Future Differ from the Past?", in *China: The Next Twenty Years of Reform and Development*, ed. Ross Garnaut, Jane Golley, and Ligang Song (Canberra: Australian National University E Press, 2010), 110–113, and ZhongXiang Zhang, *Copenhagen and Beyond: Reflections on China's Stance and Responses* (Honolulu, HI: East–West Center, June 2010). See also ZhongXiang Zhang, *Assessing China's Carbon Intensity Pledge for 2020: Stringency and Credibility Issues and Their Implications* (Honolulu, HI: East–West Center, October 2010).

97 David Cohen-Tanugi, "A Clear-Eyed Look at China's Climate Target", http://switchboard.nrdc.org/blogs/bfinamore/a_cleareyed_look_at_chinas_cli.html, posted 19 April 2010, accessed 16 March 2012. See also Deborah Seligsohn and Kelly Levin, *China's Carbon Intensity Goal: A Guide for the Perplexed* (China FAQs, World Resources Institute, April 2010).

98 David Cohen-Tanugi, *Putting it into Perspective: China's Carbon Intensity Target* (National Resources Defense Council, October 2010), 1, 2 and 8. See also, Jusen Asuka and Xiangchun Lu, *Quantified emissions reduction targets of China: Assessing the Chinese target of 40–45% reduction in* CO_2 *intensity* (Institute for Global Environmental Studies, November 2010).

99 *China and a Sustainable Future: Towards a Low Carbon Economy and Society*, 3–5, 51–59 and 99–103. The report noted that its scenarios were not predictions but "an inference" based on certain hypotheses and assumptions. It was considered "virtually impossible" to predict emission trends as they were the result of very complex and dynamic systems.

100 *Winds of Change*: *East Asia's Sustainable Energy Future* (International Bank for Reconstruction and Development, World Bank, and AusAid, May 2010), 3–7, 59, 87, and 112.
101 Jiang Kejun, *Potential Secure, Low Carbon Growth Pathways for the Chinese Economy* (Beijing: Energy Research Institute, National Development and Reform Commission, January 2011). The scenarios assumed 9 percent average GDP growth in 2005–2020.
102 Julie Chao, *A Surprise*: *China's Energy Consumption will Stabilize* (Lawrence Berkeley National Laboratory, 27 April 2011).
103 Nan Zhou, David Fridley, Michael McNeil, Nina Zheng, Jing Ke and Mark Levine, *China's Energy and Carbon Emissions Outlook to 2050* (China Energy Group, Lawrence Berkeley National Laboratory, April 2011), vii–xv, 3 and 53. See also Nan Zhou, David Fridley, Michael McNeil, Nina Zheng, Jing Ke and Mark Levine, *Peak* CO_2*? China's Emissions Trajectories to 2050* (China Energy Group, Lawrence Berkeley National Laboratory, June 2011), and Mark D. Levine, *Will China Overwhelm the World with its Greenhouse Emissions?* (Lawrence Berkeley National Laboratory, February 2010).
104 *China's Energy Technologies to 2050* (Grantham Institute for Climate Change, Imperial College London, January 2012). See also Neil Hirst, *An Assessment of China's 2020 Carbon Intensity Target* (Grantham Institute for Climate Change, August 2011), Executive Summary.
105 Nicholas Stern, *China's growth, China's cities, and the new global low-carbon industrial revolution* (Centre for Climate Change Economics and Policy (UK) and Grantham Research Institute on Climate Change and the Environment (UK), November 2010).
106 Hanns W. Maull, *Global Shift*: *The Challenges of Energy Interdependence and Climate Change* (Washington, DC: Transatlantic Academy, September 2011). By 2012 China had already reached 28 percent of global emissions.
107 *World Energy Outlook 2011*: *Executive Summary* (International Energy Agency, 2011). In its 2012 edition the IEA's updated *New Policies* scenario for China indicated annual emissions growth could slow to 1.4 percent between 2010 and 2035, but its emissions in 2035 would still be 40 percent higher. See CO_2 *Emissions from Fuel Combustion: Highlights* (International Energy Agency, 2012).
108 Chris Buckley, "China report spells out 'grim' climate change risks", *Reuters* (17 January 2012).
109 Makridakis and Taleb, "Living in a world of low levels of predictability".
110 Vaclav Smil, "Perils of Long-Range Energy Forecasting: Reflections on Looking Far Ahead", *Technological Forecasting and Social Change* 65 (2000), 251–264.
111 Li Huimin and Qi Ye, "Comparison of China's Carbon Emission Scenarios in 2050", *Advances in Climate Change Research* 2, no. 4 (2011), 193–202.
112 Two separate LBNL reports compared scenario results to 2030 and to 2050 respectively: David Fridley, Nina Zheng, and Nathaniel T. Aden, *What Can China Do?*: *China's Best Alternative Outcome for Energy Efficiency and* CO_2 *Emissions* (Lawrence Berkeley National Laboratory, July 2010), and Nina Zheng, Nan Zhou, and David Fridley, *Comparative Analysis of Modeling Studies on China's Future Energy and Emissions Outlook* (Lawrence Berkeley National Laboratory, September 2010).
113 Charles F. Doran, "Why Forecasts Fail: The Limits and Potential of Forecasting in International Relations and Economics", *International Studies Review* 1, no. 2 (Summer 1999), 11–41.
114 Smil, *China's Past, China's Future*, 210–211. He cited an example where his own forecast of China's energy demand had proved to be accurate only by chance because his *underestimation* of one factor had been cancelled out by his *overestimation* of another factor.
115 Smil, "Perils of Long-Range Energy Forecasting", 252 and 261.

116 Makridakis and Taleb, "Living in a world of low levels of predictability".
117 Spyros Makridakis, Robin M. Hogarth, and Anil Gaba, "Forecasting and uncertainty in the economic and business world", *International Journal of Forecasting* 25 (2009), 794–812.
118 Yan Yan, "China's emissions may peak as soon as 2025", *Chemistry World* (12 December 2012).
119 For example, David Fogarty and David Stanway, "China emissions study suggests climate change may be faster than thought", *Reuters* (10 June 2012). In January 2012, China also announced an absolute limit on total energy consumption by 2015, see "China's cabinet approves energy consumption control target", *Xinhua* (30 January 2013).
120 Michael Rose, "Global CO_2 emissions hit record in 2011 led by China: IEA", *Reuters* (24 May 2012).
121 However, with slower growth in 2012, carbon intensity fell by 4.3 percent. See Trevor Houser, *China's 2012 Energy Report Card*, Rhodium Group, 27 February 2013.
122 Dabo Guan, Zhu Liu, Yong Geng, Sören Lindner and Klaus Hubacek, "The gigatonne gap in China's carbon dioxide inventories", *Nature Climate Change* (10 June 2012), 1–4.
123 A Deutsche Bank report projected that, on current trends, China's PM 2.5 air pollution could increase by 70 percent by 2030, or be reduced (in ambitious reform scenarios) by about 20 percent by 2020 and over 50 percent by 2030. See Jun Ma, Michael Tong, and Audrey Shi, *China: Big bang measures to fight air pollution* (Deutsche Bank AG, 3 March 2013), 4.
124 Guy Chazan, "Shale gas boom helps slash US emissions", *Financial Times* (23 May 2011).
125 Elliot Brennan, "The 'Fracking' Revolution Comes to China", *Financial Times* (24 March 2013).
126 *China Wind Energy Development Roadmap 2050* (International Energy Agency and Energy Research Institute, 2011).
127 Smil, *China's Past, China's Future: Energy, Food and Environment*, 26 and 210–215.

Bibliography

Auffhammer, Maximilian and Richard T. Carson. *Forecasting the Path of China's CO_2 Emissions: Offsetting Kyoto – and Then Some*. Berkeley, CA: University of California, 2006.

Berrah, Noureddine, Fei Feng, Ronald Priddle and Leiping Wang. *Sustainable Energy in China: The Closing Window of Opportunity*. Washington, DC: World Bank, March 2007.

Blanford, Geoffrey J., Richard G. Richels and Thomas F. Rutherford. *Revised Emissions Growth Projections for China: Why Post-Kyoto Climate Policy Must Look East*. Harvard Kennedy School, September 2008.

Chandler, William. *Breaking the Suicide Pact: US–China Cooperation on Climate Change*. Carnegie Endowment for International Peace, March 2008.

Chandler, William and Holly Gwin. *China's Energy and Emissions: A Turning Point?* Pacific Northwest National Laboratory, 2004.

Chandler, William and Yanjia Wang. *Memo to Copenhagen: Commentary Is Misinformed – China's Commitment Is Significant*. Carnegie Endowment for International Peace, December 2009.

China and a Sustainable Future: *Towards a Low Carbon Economy and Society*. China Human Development Report, Renmin University and UNDP, 2010.

China Sustainable Development Strategy Report 2009: China's Approach Towards a Low Carbon Future. Chinese Academy of Sciences, 2009.

China's Energy Technologies to 2050. Grantham Institute for Climate Change, Imperial College London, January 2012.

China's Green Revolution: *Prioritizing Technologies to Achieve Energy and Environmental Sustainability*. McKinsey & Company, February 2009.

China's Pathway Towards a Low Carbon Economy. China Council for International Cooperation on Environment and Development, November 2009.

Delivering Low Carbon Growth: *A Guide to China's 12th Five Year Plan*. The Climate Group, March 2011.

Economy, Elizabeth C. *The River Runs Black: The Environmental Challenge to China's Future*. Ithaca, NY: Cornell University Press, 2005.

Fridley, David, Nina Zheng and Nathaniel T. Aden. *What Can China Do?: China's Best Alternative Outcome for Energy Efficiency and CO_2 Emissions*. Lawrence Berkeley National Laboratory, July 2010.

Hallding, Karl and Marie Olsson. *Balancing Climate Concerns and Energy Security: China Searching for a New Developmental Pathway*. Stockholm Environment Institute, 2010.

Heaps, Charles. *A Deep Carbon Reduction Scenario for China*. Stockholm Environment Institute, 24 May 2009.

Jiang, Kejun. *Potential Secure, Low Carbon Growth Pathways for the Chinese Economy*. Beijing: Energy Research Institute, National Development and Reform Commission, January 2011.

Levine, Mark D. *Will China Overwhelm the World with Its Greenhouse Emissions?* Lawrence Berkeley National Laboratory, February 2010.

Levine, Mark D. and Nathaniel T. Aden. *Global Carbon Emissions in the Coming Decades: The Case of China*. Lawrence Berkeley National Laboratory, 2008.

Lin, Jiang, Nan Zhou, Mark D. Levine and David Fridley. *Achieving China's Target for Energy Intensity Reduction in 2010: An Exploration of Recent Trends and Possible Future Scenarios*. Ernest Orlando Lawrence Berkeley National Laboratory, December 2006.

Smil, Vaclav. *The Bad Earth: Environmental Degradation in China*. New York: M. E. Sharpe, 1984.

Smil, Vaclav. *China's Environmental Crisis: An Inquiry into the Limits of National Development*. New York: M. E. Sharpe, 1993.

Smil, Vaclav. *China's Past, China's Future*: *Energy, Food and Environment*. New York: Routledge Curzon, 2004.

Wang, Tao and Jim Watson. *China's Energy Transition*: *Pathways for Low Carbon Development*. University of Sussex and Tyndall Centre for Climate Change Research, 2009.

Zhang, ZhongXiang. *Assessing China's Carbon Intensity Pledge for 2020*: *Stringency and Credibility Issues and Their Implications*. Honolulu, HI: East–West Center, October 2010.

Zhang, ZhongXiang. *Climate Commitments to 2050: A Roadmap for China*. Honolulu, HI: East–West Center, 4 December 2009.

Zhang, ZhongXiang. *Copenhagen and Beyond: Reflections on China's Stance and Responses*. Honolulu, HI: East–West Center, June 2010.

Zheng, Nina, Nan Zhou, and David Fridley. *Comparative Analysis of Modeling Studies on China's Future Energy and Emissions Outlook.* Lawrence Berkeley National Laboratory, September 2010.

Zhou Dadi and Mark Levine. *China's Sustainable Energy Future: Scenarios of Energy and Carbon Emissions.* Lawrence Berkeley National Laboratory, 2004.

Zhou, Nan, David Fridley, Michael McNeil, Nina Zheng, Jing Ke and Mark Levine. *China's Energy and Carbon Emissions Outlook to 2050.* China Energy Group, Lawrence Berkeley National Laboratory, April 2011.

Zhou, Nan, David Fridley, Michael McNeil, Nina Zheng, Jing Ke and Mark Levine. *Peak CO_2? China's Emissions Trajectories to 2050.* China Energy Group, Lawrence Berkeley National Laboratory, June 2011.

7 International uncertainties

It seems no overstatement to suggest that the continued growth of China's international influence, or its decline, is likely to have a decisive impact on the security of many countries and on the stability of the international system. Critical uncertainties in this domain are therefore of major global interest and concern. Both China's ascendancy or its decline could have either positive or negative effects. The highest-level critical uncertainty can perhaps be expressed simply as: *How will China's international relationships evolve?* Within this broad question, however, expert attention has universally focused on a much more specific critical uncertainty: namely, *Will China's international behaviour be competitive or cooperative?* This draws attention to the possibility of two alternative futures, one of which could potentially be enormously disruptive and damaging for other countries and also for China itself, while the other could be crucial for ensuring multilateral security. In reality it seems highly likely that the future will involve both competition and cooperation, with the key question being how far the mix will support either positive or negative outcomes. In either case, there is a great deal at stake.

In a competitive future, the specific uncertainties could include: whether China's development is likely to pose a threat to other countries; whether as its power increases its behaviour will be increasingly assertive; whether its interactions with other countries will lead to economic, military or other forms of conflict; whether, for example, its territorial claims might lead to clashes with its regional neighbours; whether China's international behaviour will lead to intense strategic competition with the United States and other major powers; whether China's military development might fuel an arms race or a new cold war; and whether China might even eventually dominate the future international system. In a cooperative future, these uncertainties could include: whether China will be a constructive player in the international system; whether its interactions with other countries will be shaped by shared concerns and a growing appreciation of interdependence; whether or to what extent China's international posture will be determined by domestic priorities; and whether improved communication and understanding may alleviate strategic mistrust and mutual misperceptions.

Within the international domain, there is an especially prodigious and constantly growing volume of literature about China's likely future attitudes and actions. As a consequence, this chapter will confine itself to highlighting the major competing views and forecasts of a selection of some of the most influential experts in other countries and in China itself. It will consider the arguments put forward that China is likely to present an increasing security and military threat, that it will be increasingly assertive or even aggressive, and that it will inevitably be a formidable strategic competitor. The chapter will also consider counter-arguments – including from China's own experts – that the threat posed by China has been exaggerated, that it emphasises peaceful development, and that it believes a cooperative international strategy is essential. This debate relies principally on trend analysis and expert judgement, but there have also been numerous instances where experts have attempted to use international relations theory to reinforce their forecasts. This theory-based analysis, and also the powerful influence of misperception and mistrust, will be considered separately.

China threat?

In the early post-Mao period, concerns among Western experts about China's future international influence were relatively subdued. The new regime led by Deng Xiaoping emphasised economic development rather than ideological fervour and appeared to tolerate gradual liberalisation. As seen in Chapter 2, perceptions took a radical turn following the Tiananmen crisis and were further heightened by the resurgence of China's economy from the early 1990s. Many assessments of China's strategic intentions in this period were extremely negative. One of the first to call attention to an alleged new China security threat was Ross Munro, who claimed that China seemed "firmly embarked on a course of explosive economic growth and military assertiveness that will indeed (as Napoleon reportedly predicted) reverberate throughout Asia and the world". It would also pose a growing threat to vital US interests, with the prospect that "an inevitably conflict-ridden relationship" might descend into "full-blown hostility". In terms that were strikingly similar to much commentary 20 years later, Munro claimed that nowhere had China been more aggressive than in the South China Sea where it was "rushing to establish an overwhelming military presence", making it "virtually certain" that it would become dominant there.[1]

An even more fully-fledged portrayal of China as a likely new threat was presented by Denny Roy. He perceived "an alarming indication" that as a potential superpower China was likely to be assertive and uncooperative. There was a strong possibility that it would use force to seek hegemony in its region, which might plunge Asia into a new cold war. As an authoritarian, unstable third-world regime and a dissatisfied irredentist power, it would be prone to militarism, hyper-nationalism and adventurism. He claimed that China was also "highly averse to interdependence, and to subjection to international norms and regimes".[2] Several other US analysts drew rather similar conclusions. Larry Wortzel saw China as seemingly locked in a "quasi-imperial competition for regional hegemony".[3] David Goodman and Gerald Segal predicted that "each time the Chinese elephant puts its foot one step further, Chinese leaders will see little reason to halt their lumbering advance".[4]

Richard Bernstein and Ross Munro, in their 1997 book *The Coming Conflict with China,* observed that many US policy-makers and experts believed that China would inevitably become Westernised, that it would be pragmatic, materialist and non-threatening. They, however, believed China's "sheer size and inherent strength, its conception of itself as a center of global civilization, and its eagerness to redeem centuries of humiliating weakness" were "propelling it toward Asian hegemony". China's military leaders accepted the need to "quietly nurse our sense of vengeance" and "conceal our abilities and bide our time", but Bernstein and Munro thought it almost indisputable that over the next one or two decades China would seek to become "the dominant power on its side of the Pacific". To this end they expected it to pursue four interrelated goals: namely, gaining sovereignty over Taiwan, achieving control of the South China Sea, inducing US military withdrawal from Asia, and subordinating Japan.[5] In the context of the 1995–1996 Taiwan Straits crisis, they considered that China had become "increasingly prickly and aggressive" and had behaved "like a rogue country". If the CCP lost its legitimacy this could lead "to paranoia and arrogance, and to the use of bluster, threat, and xenophobic appeal".[6]

Samuel Huntington's view that international competition would increasingly revolve around a "clash of civilizations" also included specific concerns about the challenge China might pose to international order. He believed its rapidly increasing military spending, its attempts to purchase an aircraft carrier, and its assertion of sovereignty over the South China Sea, were provoking a multilateral arms race.[7] He thought it would want to see East Asia "return to its traditional unipolar pattern with a hierarchy of power centered on Beijing". "China's history, culture, traditions, size, economic dynamism, and self-image", he maintained, would all impel it to seek hegemony. Like every other emerging major power it would engage in "outward expansion, assertion, and imperialism". He conjured a scenario in which a future conflict between the US and China in the South China Sea might escalate into a "global civilizational war".[8]

Several other US commentators also seemed eager to embrace extravagant depictions of China as a regional or even global threat.[9] Constantine Menges believed it would pursue a "stealthy strategy of geopolitical and economic dominance" that would put it on a collision course with the US. It believed it must "either dominate the world or be dominated by the United States and its allies". Menges detailed a remarkable scenario, with an "illustrative timeline", describing eight stages by which China would secure total global military and economic hegemony.[10] Another robust assessment was offered by Charles Krauthammer, who maintained that "any rational policy toward a rising, threatening China" should include "containing China as it tries relentlessly to expand its reach" and "undermining its pseudo-Marxist but still ruthless dictatorship".[11]

A number of Western experts based their concerns about a future China threat on its growing military capability. Morton Kaplan thought the US military contest with China would "define the twenty-first century", and that it would be "a more formidable adversary than Russia ever was". Its naval and air forces might not match the US for some decades, but it was perfecting the art of asymmetry which could allow it to achieve "a sort of political–strategic parity".[12] Jonathan Pollack, along with many other military analysts, highlighted a new Chinese emphasis on developing "anti-access" capabilities.[13] US Department of Defense reports expressed growing concern about the future effect on regional military balances of China's growing power projection capabilities. US analysts pointed to its growing submarine and surface naval forces, its evolving conventional missile strike assets, and especially its new land-based, anti-ship missile capability that would threaten US aircraft carriers.[14] Meanwhile, the PLA Navy has deployed a refurbished aircraft carrier of its own, with additional and more-capable carriers expected to appear in the future. Andrew Erickson noted that, if China used its carriers cooperatively, this would strengthen the international system, but, if it used them to intimidate its neighbours, this would be likely to lead to "great-power competition and conflict".[15]

Some drew more dramatic conclusions about the future consequences of China's military growth. Dan Blumenthal thought China was developing a "growing ability to coerce US allies, interdict US forces, and cut off US access to parts of the global commons in possible pursuit of regional hegemony".[16] John Tkacik believed China posed "a multidimensional matrix of threats", which could "blow up in America's face at a moment's notice". He highlighted its apparent intention to assert its territorial claims and to dominate East Asian sea-lanes and airspace. In particular, he believed China's "increasingly adamantine" claims in the South China Sea, the Taiwan Strait, and the East China Sea were "certain to be resolved in one of two ways: either China gets what it wants or it will use armed conflict to enforce its so-called 'core interests'".[17] Several analysts believed that previous assessments had underestimated China's ability to rapidly deploy new capabilities. Andrew Erickson and Gabe Collins observed that its anti-ship ballistic missile system had become operational "far faster than many foreign observers thought possible".[18] Amy Chang agreed that China's military modernisation had "outpaced the estimates of US and other foreign observers". As examples she cited not only its anti-ship ballistic missiles but also new Yuan-class diesel-electric attack submarines, a new anti-satellite system, and J-20 stealth fighter aircraft.[19]

Others were somewhat less concerned about China's military development. Robert Ross accepted that China was developing a new generation of military capabilities, but he believed "neither alarm nor exaggerated assessments" were justified because China had not developed the technologies to be a serious threat. The threat was "simply vastly overrated" because the US could be "confident in its ability to retain maritime dominance well into the twenty-first century". The US Navy was taking measures to minimise its vulnerabilities. China's aircraft carriers would not be a threat for decades, and in the meantime they would be simply "an additional target" for US forces.[20] A RAND report acknowledged that China was likely over the next 20 years to become "a true peer competitor" of the US, and to "achieve local superiority in its immediate neighbourhood".

But the authors considered it would remain focussed on its immediate periphery and that economic consequences would be a powerful mutual deterrent against the risk of conflict.[21]

A different perspective, but one that also tended to downgrade the China threat, was proffered by Erickson. First, he argued that, rather than continue to grow rapidly, China might "slow or even falter" due to the "increasing headwinds" it faced. It might follow an "S-curved trajectory", with accelerated development giving way to subsequent decline, greatly affecting the scale of its military challenge to the US. In any case, he assessed that China's primary challenge would be within the Yellow, East, and South China Seas and that its blue water expansion did not pose a serious problem. Even in its neighbouring waters, its forces would be highly vulnerable. Its anti-access and area-denial capabilities were significant, but he judged that even after the US moved beyond its "unipolar moment" it would "remain the world's sole super-power for years to come".[22] Referring to Minxin Pei's view that China risked being caught in a trapped political and economic transition, Erickson raised the possibility of a trapped military transition also. As the PLA attempted to integrate more advanced technologies, he thought a slowdown in the pace of capability enhancement was likely. The PLA could become "trapped in its region", especially if China's growing assertiveness led its neighbours to accelerate their own military development. It could "stir up the Near Seas" but not "make tsunamis beyond that".[23]

Another approach taken by experts to assess China's future military capabilities was to examine trends in its defence expenditure. This analysis was often hampered, however, by lack of rigour in evaluating these quantitative trends. First, the fact that China's official defence budget did not include all defence-related expenditure complicated estimation of the total budget. This led to divergent estimates of the real level of China's defence spending ranging from twice to ten times the official figure.[24] Second, estimates of expenditure trends sometimes failed to adjust for inflation. RAND noted that, at nominal prices, China's official defence budget grew eleven-fold between 1978 and 2003, but only about three-fold at constant prices. Third, analysts rarely compared growth in defence expenditure with the growth of total government expenditure. RAND noted that China's defence expenditure had remained stagnant or even declined as a percentage of total central government expenditures.[25] Fourth, there was no agreement among analysts about whether comparisons between China's defence spending and that of other countries should be considered at market exchange rates or purchasing power parity. John Tkacik argued that China's defence expenditure should be considered in terms of PPP, which "simply reflects the reality that a billion bucks buys a lot more 'bang' in China than in the United States".[26]

Others expanded on these difficulties. Dennis Blasko noted that attempts to estimate China's military budget did not use "a consistent analytical methodology" and were often guilty of "fact-free" analysis. He observed that most "reputable outside estimates" ranged from two to four times the official budget, with the Pentagon's 2010 estimate being a little over double.[27] Others noted that

China's defence spending as a proportion of GDP was still less than half that of the US, and that in absolute terms its defence spending in 2010 was only about one-seventh that of the US.[28] Adam Liff accepted that China's defence spending had become the second largest in the world, but argued that concerns about the growth of these expenditures were based on "an over-simplistic narrative" about its strategic intentions. China's military modernisation was driven by "decades of neglect and military backwardness" and increases in nominal defence spending had been "heavily mitigated by rampant inflation". Real increases had been relatively constant as a percentage of GDP and had constituted a declining percentage of total central government expenditure.[29]

Nevertheless several experts continued to express concern about China's military spending. One report commented that although it allegedly faced "no serious security threat from any of its neighbours", it was increasing its defence spending while other major powers were cutting. Moreover, the report argued, a proper comparison with the US should only consider US military spending in the Asian theatre.[30] China responded to such statements by pointing out that military expenditure in the US and other countries was also growing rapidly.[31]

Given the imponderables involved in estimating China's annual defence expenditures, it is little wonder that there has been similar uncertainty about predicting future trends, although this did not prevent some from making the attempt. *China Daily* took umbrage at (unsourced) Japanese reports suggesting that China's military expenditure could double by 2020, more than triple by 2030, and increase seven-fold by 2050.[32] However, analysts at IHS Jane's Defence reportedly expected China's defence spending to double by as early as 2015.[33] *The Economist* calculated that China's defence spending, which in 2011 was one-fifth of the US, could catch up with it by 2025.[34] Trefor Moss ventured that China could become the world's biggest military spender in the 2020s or 2030s, although he wisely added that there were too many unknown variables to make an accurate prediction.[35]

Assertive China?

As the new century progressed, it became less common for experts to portray China unequivocally as a threat. Instead, there was a tendency for earlier comments about an "increasingly assertive" China to re-emerge. This formulation had the advantage of leaving judgements about China's future intentions more open while still raising concern about the direction of its international strategy. In 2000, Michael Swaine considered three possible scenarios – a *Chaotic China*, a *Cooperative China*, and an *Assertive China*. He considered that there were "extremely powerful" reasons to expect an assertive strategy "for all the time-honored reasons associated with power-politics".[36] Alistair Johnston was more cautious, asking whether China was currently pursuing a status quo foreign policy or whether it was "a dissatisfied, revisionist state". He concluded that it was not clear whether this revisionist label had "at this moment in history" become appropriate but could not guarantee it would not become so in the future, especially in the event of domestic upheavals or "a spiralling China–US security dilemma".[37]

By the mid-2000s, Minxin Pei concluded that China's international posture was predisposed to be assertive, under the influence of its authoritarian political system, its realist worldview, and Chinese nationalism. These three "demons" would only be moderated by an "assertive pragmatism" which would lead China "occasionally" to behave like an internationalist power, even though this did not reflect its true disposition.[38] A similar view was maintained by John Lee who argued that, for pragmatic reasons, China avoided overt confrontation with the West, but in fact it saw the United States as a "strategic competitor". It behaved as a "largely cooperative" power, but was also "a disruptive and subversive one".[39]

Following the GFC, and especially from 2010, many Western commentators perceived important changes in China's approach to international affairs due to increased confidence in its status and influence. They pointed to greater assertiveness regarding US arms sales to Taiwan and to a more active posture regarding its maritime claims in the East and South China Seas. Minxin Pei asked whether an assertive China had become the "New Normal". He believed its growing tensions with the West stemmed from powerful dynamics that would lead to "repeated disagreements or even acrimonious confrontation". Clearly, he claimed, China was no longer keeping a low profile as counselled by Deng Xiaoping, but was "flaunting its newly acquired power and status" to its domestic audience.[40] Gordon Chang also perceived an "increasingly assertive" China, with hawkish officers in the PLA becoming more influential. He assessed that "China's ambitions know few bounds these days".[41]

A special report from *The Economist*, with input from many experts, raised the question of whether China was "Friend or Foe?" Its "keen sense of its growing national power and American decline" had reportedly begun to alarm China watchers. One Chinese scholar was said to have urged China to "give up the illusion of US partnership and face squarely the profound and inevitable strategic competition". The report concluded that the "most likely outcome" was "a more assertive China that wants to get more done abroad without fundamentally upsetting the world order". On territorial issues it might be "uncompromising and increasingly unreasonable", although on other issues it would probably be prepared to negotiate.[42]

David Shambaugh also expressed considerable concern. After spending a year in China in 2009–2010, he judged that the world had progressed to a new phase of Chinese assertiveness. He reportedly described China's diplomacy as "very truculent, sometimes strident, sometimes arrogant, always difficult".[43] He expected that this would be remembered as the period when China became "more difficult for the world to deal with". He wondered whether this would be temporary, because it represented mainly the views of conservative and nationalist elements that were only one part of "an extensive, animated, and diverse domestic debate". He concluded that China had "multiple international identities and a schizophrenic personality" but the centre of gravity was "anchored" by realists. This portended an increasingly self-interested nation that would seek to maximise its own power. China's behaviour would often appear inconsistent and conflicted, but with "a predominant realist, and troubling, character".[44]

Competitive China?

By the end of the first decade of this century, perceptions of China's increased assertiveness and of its growing military strength stimulated a wide-ranging dialogue among experts about whether its relationships with the United States and with other regional countries were likely to become increasingly competitive, or whether – under the continued influence of a cautious and pragmatic policy – they would remain mainly cooperative. What was noteworthy about this discussion was that a growing number of experts, both in the West and in China, appeared to believe that the element of competition was growing and the impetus for cooperation declining. In the West, some experts appeared to reach a deterministic conclusion that China's rise would be accompanied by an intensifying strategic competition with other major powers, and especially with the US. Nick Bisley, for example, considered that China's "distinctly more assertive tone" during 2010 was unlikely to be an exception following which it would "go back to hiding its claws and getting on with domestic development". Rather, he saw China's new confrontational posture as "a harbinger of a new, evolving regional order in which friction and conflict among Asia's major powers will increasingly become the norm". The region was likely to be increasingly unstable, with the potential for events to escalate "in unsettling and potentially destructive ways".[45]

This revitalised notion of an inevitable and destabilising strategic competition in Asia, with China at its epicentre, became increasingly popular in some quarters. Its most committed contemporary spokesperson is perhaps Aaron Friedberg. In his 2011 book *A Contest for Supremacy*, Friedberg stated that he had long been troubled by a "wilful, blinkered optimism" about the prospects for US–China relations. Most China experts, he thought, seemed to believe rivalry between the two countries was "highly unlikely, too terrifying to contemplate". But he noted that concerns about this rivalry had now become more widespread. He believed that the two sides had become "locked in a quiet but increasingly intense struggle for power and influence". The US aimed to promote regime change in China, nudging it gradually toward liberal democracy. Beijing for its part felt the need to constrict US influence in the Western Pacific "pushing it back and ultimately displacing it as the preponderant power in East Asia". He asserted that, for more than a century, the US had not faced a strategic competitor such as China, which had more people, resources, and industrial strength than Nazi Germany or imperial Japan, and a more efficient economy than the Soviet Union.[46]

Friedberg thought optimism about China could prove to be "profoundly dangerous". The regional military balance would "start to tip sharply in China's favour" and the US could find itself "pushed to the margins of Asia if not out of the region altogether", compelling it to acknowledge China as the preponderant regional power and leaving China free to enforce its territorial claims. Much commentary on China, he said, was based either on "straightline-itis", which assumed prevailing trends would continue, or "collapse-ism", which assumed what goes up must come down. He judged, however, that "for the moment" China's continued rise was "a better bet than catastrophic decline". In these circumstances, his conditional prediction was that US–China relations would become "increasingly tense and competitive". Cooperation might persist, but it was likely to decline as the relationship became "increasingly brittle". He already saw evidence of "an increasingly intense and focussed military competition" and judged that US power projection forces in East Asia could become "wasting assets" unless it made substantial new investments. He suggested several ambitious capability enhancements, including being able to threaten China's economy and perhaps also its mechanisms for domestic political control.[47]

In a later article, Friedberg's concerns seemed to have strengthened. He thought the bilateral power balance was "set to shift sharply in Beijing's favour". China's ruling elites had become "both arrogant and insecure", due to their desire for regional hegemony and Washington's unwillingness to allow it. China's "truculence" was causing its regional neighbours to welcome the "more muscular rhetoric" coming from Washington.[48] John Lee supported Friedberg's strategic assessment. He thought China's ascent was "likely to spark an intense security competition with the US, leading to the strong possibility of war". China was "a revisionist power" due to its "fundamental dissatisfaction with several of its land and maritime borders" and its "desperation to secure its 'core interests' ". China's "strategic dissatisfaction and impatience" could lead to "disastrous miscalculation and error".[49]

Although David Shambaugh did not favour the same solutions as Friedberg, he also continued, in 2013, to be pessimistic about prospects for the bilateral relationship. Comparing Chinese perceptions of the US during his stay in Beijing with his impressions from earlier visits over thirty-two consecutive years, he concluded that "something very negative" was happening. His contacts appeared to regard the US with "a combination of disinterest and disgust". A "new robust Chinese nationalism" was on display. On his return to the US he found this negativity was reciprocated in a "paradigm shift" among Washington's China watchers. Shambaugh saw China and the US as "tangled titans", with numerous interdependent interactions in many domains. They were "inextricably tied together" and cooperated extensively, but he perceived a "rising competition". He labelled this situation as "coopetition". Competition was becoming "primary" and cooperation "secondary and declining". The tangled titans would find it "increasingly difficult to co-exist – yet they must".[50]

Meanwhile a European perspective was provided by Jonathan Holslag, who feared that the threshold for China's use of force in Asia would "almost certainly" be lowered by its growing military strength, augmented by its growing confidence and the pressures of nationalism. He predicted that in the 2010s China would "almost certainly" shift its military build-up "into a higher gear", although it would need to balance the benefits of greater capability against the threat to its interests if it conveyed "hostile intent". But "in the end" it would not be how China behaved that would matter but how other states perceived it "*could* behave in the future". This meant, he argued, that worst-case scenarios would then "ultimately be the likeliest bet".[51]

Peaceful China?

Confident depictions of a China threat, of intent to use its increasingly capable military forces to project its power, and of an assertive/aggressive/competitive strategic posture were accompanied by a variety of counter-arguments from other experts during the post-Tiananmen period. Even Richard Bernstein expressed concern that the discourse on China had become "radicalized". When he and Ross Munro wrote their 1997 book, they believed it would "nail down one end of the emerging spectrum of views on China", but they now found themselves in a middle position between those who advocated engagement and those for whom China was "a new incarnation of the old Soviet expansionist menace". He did not believe China had the ambition or the resources to dominate the world.[52]

David Lampton also rejected the proposition that China was "a rogue state with hegemonic ambitions". It had made "concerted and successful efforts" to improve its relations with all neighbouring countries, participated constructively in many international institutions, and adhered to international treaties and regimes. Its military capabilities were modest, it was not undertaking "a huge military build-up", and its inflation-adjusted defence spending was "much less significant" than others claimed.[53] Ted Carpenter and James Dorn also assessed that, rather than attempting to "bully its neighbours and dominate Asia", China was more likely to be a stabilising and cooperative power. Unlike Nazi Germany or the USSR, it was not "a messianic, expansionist power" but a normal rising power that did not pose "a malignant security threat".[54]

Similarly, Michael Swaine was disturbed by a trend in China watching that perceived a belligerent China as "the primary threat to peace in Asia", engaging in a "crash program of military expansion" designed to coerce its neighbours and eject the US from the region. He said that such analysis was "fundamentally wrong". Its proponents routinely employed "distortions, half-truths and, in some cases, complete falsehoods". In reality, he said, China's military upgrade sought to remedy "a hopelessly obsolete … and largely ineffective army", and it was subordinate to sustaining economic growth. He agreed that China's growing military power could not be viewed with indifference, but it could not be accurately assessed using the "faulty data, political rhetoric and simplistic assumptions" of China threat proponents.[55]

In a survey of competing views in the United States about the alleged China threat, Joseph Nye in 1997 noted that, on the one hand, Arthur Waldron believed war in Asia was "probable", Robert Kagan thought the Chinese leadership saw the world "in much the same way Kaiser Wilhelm II did a century ago", and *New Republic* magazine believed China had "embarked on a program of massive militarization" and was on track to became "the other military superpower on earth". On the other hand, Nye found that many analysts were sceptical about a Chinese threat in terms of either its capability or intentions. Owen Harries thought vested interests in the US were "suffering from an enemy-deprivation syndrome"; Andrew Nathan and Robert Ross believed that, for the foreseeable future, China could not "establish regional domination or become a global superpower"; and David Shambaugh thought "recent hype" about a China threat was "grossly overblown, not empirically grounded, irresponsible and politically dangerous". For his own part, Nye conceded that, if China was unconstrained, it might wish to expel the US from East Asia and exercise hegemony, but "in the real world" states learned to be practical. He therefore assessed that prospects for avoiding conflict with China were good, provided the US disregarded "the advice of pessimists who would seize the moment to turn a vision of inevitable conflict into a self-fulfilling prophecy".[56]

During the 2000s, other experts continued to reject the China Threat hypothesis in its more extreme forms. Herbert Yee concluded that many such assertions were "exaggerated or simply groundless, and are full of bias, prejudices, and even racist accusations". He considered that "mainstream thinking" among Chinese leaders and academics favoured a stable long-term relationship with the US.[57] Contrary to the views of "anti-China hands", Emma Broomfield also thought its portrayal as a looming military threat was "a gross exaggeration".[58] David Lampton considered that China was emphasising economic rather than military growth because the latter would alarm the rest of the world, produce "countervailing coalitions", and divert resources from pressing domestic priorities.[59] He drew attention to China's "vulnerabilities and uncertainties". It would not become another Soviet Union because it had shown little inclination to build "an excessive strategic arsenal". Lampton believed it would be "a strategic blunder of breathtaking and historic proportions" if US policy managed to transform a country focused on its internal development into an adversary. He also thought that China's leaders increasingly understood that they had an interest in assisting the development of a stable international system.[60]

Lampton's views were echoed by Harry Harding, who considered it unlikely that China's rise would lead to military conflict unless its "core interests" were threatened. It did not have fundamental concerns about an international system that had benefited it so much, and its economic interdependence would deter it from military adventures.[61] Even Minxin Pei did not see China as an expansionist power. He considered that it was hemmed in by powerful regional rivals and therefore would not become a dominant hegemon in Asia, let alone further abroad. It might flex its muscles on territorial and economic issues, but would draw the line at empire building overseas. China, he concluded, might want to be a superpower but "it is not, and will not be one".[62]

Andrew Nathan was also doubtful about China's future impact. While noting, in 2009, that it would be the world's largest economy "within the next 30 or 40 years", he suggested it would be "a very new kind of number one" because on a per capita basis it would still be poor. Its focus would be internal and it would remain only "a supporting player on the global economic stage". Its ascension could be "derailed by domestic instability or global economic shocks". It would not dominate militarily because there would be "no power vacuum to expand into" and it would not be able to "force its way to supremacy". He concluded: "Neither conquest nor domination is an option." Looking to the longer term, he asked if China might achieve superpower status, but he judged that this would be "very far into the future".[63]

A thought-provoking analysis of China's response to "China threat theories" was presented by Yong Deng. Chinese analysts, he commented, saw these theories as based on an exaggerated and anachronistic view of its strategy. They nevertheless perceived the need for "behavioural adjustments" that would disprove the theories and build a new image of a cooperative and responsible China. They therefore promoted the concept of China's peaceful rise, hoping that the shared challenges of globalisation would "transform great-power politics". Deng saw China as a victim of the phenomenon of "categorisation", in which its "out-group" status meant it was very difficult to convince the US of its benign intentions.[64]

With regard to other criticisms about China's excessive assertiveness, Michael Swaine conceded that such behaviour had become a persistent new theme and that its cautious, low-profile image had given way to that of a "more confident, assertive (some say arrogant), anti-status quo power". He pointed out, however, that Western and Chinese observers saw this development from very different perspectives. Western commentary often portrayed China's assertiveness as brash, insulting, anti-Western, driven by chauvinism and strident nationalism, and posing a challenge to international order and security. But Chinese sources viewed its actions as defensive and they highlighted its limited capabilities, its policy of peaceful development and its opposition to hegemonism. Swaine believed that the majority supported non-aggressive and cooperative interaction with other countries.[65] Thomas Christensen also questioned the extent of China's assertiveness. He agreed that its new posture was more truculent, hypersensitive, abrasive, and even more arrogant than previously, and "a world apart" from its previous strategy of reassurance and emphasis on peaceful rise. But he too thought this was "better understood as reactive and conservative rather than assertive and innovative". He suggested that other countries should try to channel the aspirations of a "constructive and assertive" China into helping to address pressing global challenges.[66]

Cooperative China?

Aaron Friedberg's argument that China's international strategy would inevitably be strongly competitive and disruptive drew a variety of counter-arguments. His pessimistic views were contested by other experts, who continued to expect that with appropriate diplomacy China's impact on world affairs would be less disruptive and its behaviour more cooperative. Andrew Nathan, expanding on his previous views, thought Friedberg's attempt to synthesise the views of China's hawks had created "a false unity". He countered that China's policy-makers had "consistently been more cautious in practice than the Chinese media have been in their rhetoric". A serious audit of China's capabilities would show it was "bogged down both internally and in Asia generally". China's leaders, he considered, acted "more like people under siege than like people on an expansionist warpath". Given its many limitations, he argued that, even if China stayed on course, it could not hope for "anything that can be called supremacy, or even regional mastery" unless US power declined radically. Furthermore, "the more China rises, the more most of China's neighbours will want to balance with the US, not against it".[67] Writing with Andrew Scobell, Nathan emphasised the continued limitations of China's "overstretched military".[68] Similarly, Lyle Goldstein argued that China was rising "in a thicket of strong-willed and suspicious competitors in Asia" and that "natural balancing processes" would constrain its ambitions.[69]

An alternative to Aaron Friedberg's pessimistic assessment was provided by Michael Swaine who re-entered the debate with his own book *America's Challenge*. He saw the potential for strategic competition to be managed successfully and recommended that the US should continue to emphasise a dual policy of "cooperative engagement balanced with hedging", because China was likely to be "neither a sworn enemy or an ally or close and trusted friend". He was concerned that growing mutual distrust was already producing "a dangerous zero-sum mind-set" in the defence communities of both sides. He urged Washington to "push back against hawkish voices" and examine long-term alternatives to the "current emphasis on US military preponderance". He noted that China had defined a set of core interests as a "marker" for other countries, but like the US it favoured a strategy that combined both cooperation and hedging and sought to balance the influence of "hard-liners and more moderate elements". He believed that the two countries faced "a growing number of common problems in this new century that demand cooperative solutions", which would "reinforce incentives … to avoid worst case outcomes". Swaine saw "very little, if any, evidence" that China would consider altering its strategy of peaceful development "during the next few years". He recommended working toward "some kind of cooperative balance" among major powers in the region, or a variant of a "concert of powers" although this might be problematic.[70]

Henry Kissinger has been especially prominent among those who are hopeful about the prospects for a cooperative relationship between the US and China. He declared that the assumption that strategic confrontation with China was inevitable was "as dangerous as it is wrong", and that military imperialism and provocative diplomacy was "not the Chinese style". He considered that, when China expressed its cooperative intentions, this was "less a preference than the strategic realities".[71] Kissinger noted that groups in both the US and China now claimed a contest for supremacy was inevitable and "perhaps already under way", making appeals for cooperation seem "outmoded and even naïve". He countered that Americans should remember that even when China's GDP equalled that of the US, it would still be preoccupied with huge domestic challenges.[72] In his book

On China Kissinger suggested a future US–China relationship based on "co-evolution" or a "Pacific Community" in which both sides would seek to cooperate where possible and minimise conflict.[73] Joseph Nye also cautioned against assuming that conflict was inevitable. It would be a mistake, he argued, to base fear of China on historical analogies – for example, comparing China to imperial Germany. The facts, he concluded, did not currently justify "alarmist predictions of a coming war". There was "time to manage a cooperative relationship".[74]

A new perspective from outside the United States was provided by Hugh White, whose assessment of the likelihood of competition in the emerging US–China relationship seemed at first sight pessimistic. In early articles he speculated about the likelihood that China's rise would have a major, and probably negative, influence on the geopolitics of the Asian region. Optimists who thought it could be "accommodated within a unipolar global order under US leadership" might be right, but he thought this inherently unlikely as US primacy eroded. His main conclusion, however, was that the most damaging consequences of strategic competition could potentially be avoided. As a "second-best outcome", short of US primacy, he recommended a new power-sharing arrangement – in the form of a "concert of powers" in Asia. Nevertheless, he saw China and the US as "strategic competitors, pure and simple". Moving to a new stage of "cautious but respectful cooperation" would not be easy, especially for US realists who believed China was determined to establish its own hegemony in Asia. White thought the "most likely outcome" of such a negative mindset would be the emergence of "a systematically adversarial" US–China relationship – something that seemed to be viewed with "surprising equanimity" by some in the US.[75]

Following the GFC, White saw further evidence of China's geo-strategic ascendency. He estimated that it would become the world's richest country around 2030, which would make it "too strong to live under American leadership in Asia" and cause it to seek to "lead in its own right". He thought, however, that China would seek a balance between "maximising its influence while minimising disorder". The least likely possibility was that it would attempt to "impose on Asia a harsh hegemony", although there was "always a risk" that future leaders would become more ambitious as China became stronger. He again suggested a "concert of Asia" in which power would be shared in a collective leadership with other regional powers. This would only be possible if the US allowed China "some political and strategic space". If it did not do so, the US risked "drifting into strategic rivalry" without weighing the great costs involved. In White's view China was already too strong to be contained "without intense and protracted pressure from America" that would risk "major war – even nuclear war". Yet the "drift to antagonism" had already begun. China's leaders had become "more pushy" making the US "all the more inclined to push back".[76]

White spelt out his concert of Asia proposal in greater detail in his book *The China Choice*. He believed the future would be "bleak" without an accommodation between China and the US. The relationship would "always be competitive", but the question was whether the two sides could nevertheless continue to cooperate. They were "already taking the first steps" toward a relationship that precluded cooperation. The US, he suggested, had three options: to "resist China's challenge and try to preserve the status quo in Asia", to "step back from its dominant role in Asia, leaving China to establish hegemony", or to allow China "a larger role" in the region while "maintaining a strong presence of its own". Most Americans, he said, assumed the first option was the only choice; few considered the second; and most did not consider the third option, which he argued would best serve US and Chinese interests. Any attempt by either to dominate would lead to "sustained and bitter strategic rivalry, imposing huge economic costs and a real risk of catastrophic war".[77]

A similar view was adopted by Australia's former Prime Minister Kevin Rudd, who also emphasised the critical importance of avoiding major confrontation between China and the US. This, he thought, was "difficult but doable" within a relationship that accepted "the reality of competition, the importance of cooperation, and the fact that these are not mutually exclusive propositions". China underestimated, in his opinion, the uncertainty created by its pursuit of its territorial interests and the security dilemma this created. Like White, Rudd thought a US policy that either accelerated strategic competition or sought to maintain the status quo would be much less appropriate than a "third possibility" that recognised the reality of strategic competition but built strategic cooperation through success in specific projects. Presidents Obama and Xi had "a unique window of opportunity" to disprove the "deeply depressing historical norm" that the rise of new great powers would trigger global conflict.[78]

Chinese perspectives

At the same time as experts outside China debated whether it would be driven by a desire to compete with the United States and promote its "core interests" or whether its role would be benign and constructive, an internal debate evolved within China that reflected not only reaction to this external criticism but also divergent ideas about the most appropriate response. First and foremost, however, it should be noted that the Chinese government moved quite early to place its pursuit of a peaceful and cooperative approach at the centre of its officially declared foreign policy, and has insisted repeatedly and consistently that it will continue to adhere to this policy. The internal debate has therefore been strongly influenced and shaped by this enduring policy stance, even though individual experts have questioned some of its assumptions and the government sometimes acted in ways that weakened its credibility.

China traces this policy to the Five Principles of Peaceful Coexistence first agreed between Zhou Enlai and Jawaharlal Nehru in 1954. The concept of "peaceful rise" was popularised in the early 2000s by Zheng Bijian – a senior Chinese scholar and former official – who emphasised that China would not "seek hegemony or predominance" and explicitly welcomed a "positive" US role in the Asian region.[79] "Peaceful rise (*heping jueqi*)" morphed into "peaceful development (*heping fazhan*)" – a new policy initiative that was explained at length in a 2005 government white paper. But despite its continued fervent assertions affirming peaceful development and rejecting expansionism and hegemonism, China experienced considerable difficulty in convincing the rest of the world of its peaceful intentions. In 2011 it issued a second white paper about peaceful development that "declared solemnly again" that this was China's "strategic choice". It would not pose a military threat to any other country, and was committed to solving disputes and crises by peaceful means. It saw peaceful development as a global trend necessitated by a growing list of increasingly severe global challenges and common security issues that had to be handled cooperatively.

China's international relations scholars generally supported this official viewpoint, or were at least wary of challenging it directly. During the mid-2000s, Wang Jisi championed a moderate, prudent and cooperative international stance for China. His opinions appeared to represent the dominant viewpoint among scholars at that time. He thought China's leadership was "very cautious and sober-minded in its projection of Chinese power", and that its actual behaviour confirmed this. It was a status quo power rather than a challenger to the current world order. He believed it was "certain" that China would continue to be "defensive and prudent". The Chinese found it hard to understand assertions that their country was a threat to the outside world. They viewed such portrayals as "international schemes and conspiracies".[80] Most Chinese also considered that US primacy would eventually decline, but he believed its dominance would continue for a long time.[81] Wang emphasised the strength of US "hard power" and stated that strategic planning in Beijing was "not based on an assessment of a weakened US at present or any time soon". Therefore "leading Chinese strategists" favoured a low international profile, a moderate foreign policy, and not directly challenging the US.[82] Other scholars at Peking University supported this moderate stance. Jia Qingguo thought the peaceful development concept would reassure the world of "the benign nature of China's rise".[83] Zhu Feng commented that China would be constrained by US unipolarity. It would "avoid military provocations and foreign policy adventurism", provided the US accommodated its core national interests.[84]

Post-GFC, however, there were clearly important shifts in the perceptions of many of China's international relations experts about its most appropriate international posture. Most favoured a more proactive role. Some tended to take a distinctly negative view of US policy and actions, and were pessimistic about the prospects for bilateral cooperation. Andrew Nathan and Andrew Scobell assessed that China's experts regarded the US itself as "a revisionist power that sought to curtail China's political influence and harm China's interests". The US was "omnipresent" for China, which felt "an acute sense of its own vulnerability". Chinese analysts saw their country as "pacifistic, defense-minded, non-expansionist, and ethical" while the US was "militaristic, offense-minded, expansionist, and selfish". A small group of Chinese analysts saw US and Chinese interests as "not totally at odds", but Nathan and Scobell believed these were "outnumbered" by others, especially from China's military and security agencies, who thought it must stand up to the US.[85]

In this new situation, former mainstream strategists like Wang Jisi found themselves fighting a rear-guard action. Wang noted that China had redefined the purpose of its foreign policy to emphasise safeguarding three core interests, namely its political stability, sovereignty and territorial integrity, and economic and social development. He criticised "reckless" and unauthorised statements that had identified specific core interests beyond the issue of Taiwan. He also criticised "popular" but "misguided" attempts to portray the United States as a major threat. To base China's grand strategy on this notion, he argued, would "seriously hold back" its economic development.[86] He appeared concerned that China's foreign policy had become "more proactive than ever before", putting so much emphasis on safeguarding core national interests that it was perceived internationally as "increasingly assertive".[87] Wang was again supported by other scholars like Zhu Feng, who considered that China was still "an adolescent power" and its power disparity with the US had not "significantly narrowed". He assessed that both sides wanted "to handle the conflict, not escalate it".[88]

In the post-GFC era, however, the opinions of China's moderates continued to be contested by those who apparently favoured a more robust foreign policy. An unusually candid example of such views was provided by Lin Limin at the China Institutes for Contemporary International Relations, one of Beijing's most influential think tanks. He saw a new international strategic situation characterised by the decline of the West and the rise of non-Western countries. He accepted that the US would nonetheless retain its status as the only superpower for a long time, and that even if it lost this status it would still be "a first-rate world power". But, apart from the US, China was "like a triton among the minnows" compared to other powers such as Russia, Japan and Europe. He suggested that, if China wished to become a world power, it could begin by first becoming the leading Asia–Pacific power. In order to "dominate the world's agenda", China would need to "first pursue geo-political dominance in the Asia–Pacific". Such a strategy would be easier than attempting a "direct ascension to global power". Lin noted that, after the Cold War, people began to believe peace and development were the theme of the times and that a new era of coexistence was emerging. He asked, however, how the US was likely to react as China progressed through its "strategic sprint" phase and "accelerated to the position of global 'boss'" after it overtook the US economy in about 2020. Would the West accept China's rise peacefully or would it consider "a preemptive war"? He suggested that China would need to adopt a prudent strategy that avoided pushing other Asian countries to align with the US, but it should also enhance its military strength and "make spiritual preparation for a potentially devastating war".[89]

Yan Xuetong at Tsinghua University is one of China's most controversial international relations experts. He is often described as a hawk, although his judgements are perhaps more subtle. He has expressed many novel and not always politically correct views such as that the trend toward multipolarity is fading as the world transitions from unipolarity toward bipolarity, with China becoming a "less powerful counterpart" to the US. In contrast with China's usual disparagement of alliance arrangements as examples of cold war thinking, he has stated that a shift away from unipolarity would not be possible unless China made "a formal military alliance" with Russia.[90] He argued nonetheless that the rise of China was unlikely to lead to conflict, and that the only kind of hegemony it should seek was a "humane hegemony" that would result when it received "more international support than others".[91] He sought to explain fluctuations in US–China relations through use of a "theory of superficial friendship". In his opinion a policy of "pretending to be friends" had led to an unstable and ineffective relationship with the US, because it did not recognise that they had "fewer mutually favourable interests than unfavourable ones". He suggested that it might be better to transform the relationship "from superficial friendship to superficial enmity", which would provide a better basis for addressing conflicting interests through "positive competition and preventative cooperation".[92] Yan considered that China and the US were "moving towards all-out competition" but across a broader range of both military and non-military areas than had been the case between the US and Soviet Union. He has likened this competition to "a game of football" rather than "a boxing match", in the sense that it could lead to "occasional conflicts" but violence would not be "the primary means".[93]

In his 2013 book cited in Chapter 2, the *Inertia of History*, Yan asserted that inertial forces were causing China's rise to continue and that by 2023 it would be on an equal footing with the United States as a superpower. There would be a bipolar system in which both China and the US would take the lead. Unlike the Cold War with the Soviet Union, however, there was zero chance of a "hot" war. He believed that "peaceful competition" would be essential. But China would need to break through the barriers of its non-alignment policy and its low-profile diplomacy. It would need "true allies" to avoid being isolated and to empower it to achieve a more equal international order.[94] Yan has also suggested that by promoting the Chinese values of "benevolence, righteousness and rite", rather than the American values of "equality, democracy and freedom", China can "not only transcend the United States materially but also in terms of values".[95]

A helpful guide to competing views in China was given by Wu Xinbo from Shanghai's Fudan University. He observed that Chinese academic circles had become "more sharply divided" on the nature of the US–China relationship. Optimists, including himself, argued that the two countries were "more likely than not to become real partners", while pessimists contended that structural contradiction and conflicting interests would make it difficult to develop "real strategic friendship". He offered three alternative scenarios for the future of US–China relations in which the two countries might become partners, competitors, or enemies and assessed the most likely relationship would be "a blend of cooperation, competition, and limited conflict". China and the US would be "both partners and competitors … but not enemies". He also anticipated that China's growing self-confidence would make it "more active in agenda-setting and tougher in bargaining".[96]

Wu appeared to support this tougher stance. He noted that China's activism regarding its territorial claims in the East and South China Seas had created concern in other countries that its foreign policy had taken a disturbing new direction. However, he and other Chinese scholars have argued that this activism was reactive to "the so-called US pivot", and that China had been forced to respond when other Asian countries took advantage of the pivot to push their own territorial claims. Instead of backing down, China's response had been to adopt a "tit-for-tat" strategy. He warned that if the US continued to attempt to counter-balance China's rising influence, heightened competition and regional instability would be "inevitable". He considered that US interests would be better served by working towards a pattern of "shared power and responsibility" in the region.[97]

Side-by-side, however, with China's forceful assertion of its determination to protect its core national interests, a new and parallel emphasis on cooperative relationships has re-emerged. Following Xi Jinping's February 2012 visit to the United States there were frequent references to China's desire to develop a "new type of major power relationship". Many of the same Chinese scholars who emphasised the reality of competition simultaneously highlighted the necessity for cooperation with the US and other regional countries. They set important boundaries for China's ambitions. For example, Lin Limin re-emphasised that China should "adopt a modest and prudent strategy" that expanded cooperation and limited the risk of conflict.[98] Regarding China's maritime territorial claims, he recommended that "putting aside disputes and working at negotiations should remain our consistent stance", and that a "small group of trouble-makers" should not be allowed to "lead us by the nose over a couple of tiny cliff islands". China could not achieve "strategic ascendancy and permanent stability and security" in an antagonistic Asia–Pacific environment.[99] Lin advocated that China should choose a "cooperation paradigm" in preference to a "conflict paradigm". If it chose cooperation it could "secure the best interests at the lowest risk", while if it chose conflict it would either win "a tragic victory" or be defeated. He perceived that although there was widespread pessimism about US–China relations, leaders and decision-makers in the two countries had a deeper understanding and displayed "a cautious optimism". He was encouraged that the US had begun to engage China on a more equal footing and that its "habitual arrogance and domineering attitude" had been "dampened significantly".[100]

The majority of China's international relations experts seem united in supporting a continued emphasis on cooperative relationships, and in reaffirming its clear and long-standing official policy of peaceful development. To a greater or lesser degree, however, they – like their government – also continued to stress China's determination to uphold its core interests and associated requirements. These include its expansive and seemingly inflexible maritime territorial claims, its opposition to US arms sales to Taiwan, and its concerns about strengthened US alliance relationships. In the words of Rear Admiral Yang Yi, a former director in China's National Defense University, peaceful development is "an important part of China's strategy", but this did not mean "that China can stand humiliation and provocation".[101] Such comments indicate that in the eyes of China's experts there is real scope – if not an imperative – for its future international relations to develop cooperatively, but still considerable risk of antagonism and conflict with its neighbours over specific issues where China evidently considers it has little or no room for compromise.

The utility of theory

This chapter has so far avoided – with some difficulty – the much-discussed issue of whether, or to what extent, predictive theory may assist experts to forecast the likely trend of China's international relations. If it could be shown theoretically that China's international behaviour is predetermined – or at least strongly influenced – by factors inherent in the nature of the international system, this might give greater weight to the views of those experts who based their opinions on such theory. External perceptions of modern China have been especially influenced by one particular school of Western international relations theory, namely "realism". Realism, and especially structural realism, attributes conflict among states to the structural anarchy of the international system rather than to the characteristics of individual great powers.

One of the major exponents of structural realism has been John Mearsheimer, who also argued the case for "offensive realism" which maintains that states are driven to promote their interests by maximising their share of world power.[102] His view about China's likely negative international impact was forthright. In his *The Tragedy of Great Power Politics*, he commented that China had the potential to become "considerably more powerful" than the United States and that it would "almost certainly use its wealth to build a mighty military machine". It would be "more powerful and dangerous" than Wilhelmine Germany, imperial Japan, Nazi Germany or the Soviet Union. If China became "a giant Hong Kong" it might have four times the power of the US and "gain a decisive military advantage". It would be "strongly inclined" to seek hegemony, not because it had "wicked motives" but because "the best way for any state to maximize its prospects for survival is to be the hegemon in its region of the world".[103]

Mearsheimer later elaborated that China's rise could not be peaceful because its continued economic growth would lead to "an intense security competition" with the US. To predict Asia's future, he said, "one needs a theory of international politics that explains how rising powers are likely to act and how other states in the system will react to them". His theory said that the structure of the international system "forces states concerned about their security to compete with each other for power". China, therefore, would "try to dominate Asia". It would recognise, in Mearsheimer's colourful metaphor, that it was "better to be Godzilla than Bambi". He doubted China would "go on a rampage and conquer other Asian countries", although this was "always possible". But it was likely to want to "dictate the boundaries of acceptable behaviour" to neighbouring countries, and try to "push the United States out of Asia". The problem, he argued, was that the US for its part "does not tolerate peer competitors" and it would "go to great lengths to contain China and ultimately weaken it". Mearsheimer regarded this prospect as "categorically depressing" and wished he could be more optimistic, but international politics was "a nasty and dangerous business".[104]

Mearsheimer later criticised those who trusted China's protestations about its peaceful intentions, noting that these could not be "empirically verified". Its military capability would "change markedly" and: "Then, we will see how committed it is to the status quo." China's leaders also could not be sure of future US intentions and were likely to see it as "a war-like and dangerous country". He conceded that there was a high degree of economic interdependence between the US and China but doubted that this would prevent strategic competition. In two decades, China's military would rival the US. Moreover, the US would need to project its power across the Pacific, while China could operate "in its own backyard".[105] In an interview in China, Mearsheimer accepted that for several years there would be "powerful incentives" for China and the US to manage potential conflicts, but he insisted that a stronger China would "take greater risks".[106]

Another realist, Robert Kaplan, observed that social science theories were ultimately "gross simplifications" and that even the "most brilliant of them" would be right only about 75 percent of the time. He nevertheless found Mearsheimer's theory-based argument to be compelling. Mearsheimer had told him he was constantly meeting people who were impressed by China's desire for a peaceful relationship with the US, but he would reply that "these Chinese will not be in power in 20 or 30 years, when circumstances may be very different". He added: "Because we cannot know the future, all we have to rely upon is theory. If a theory can explain the past in many instances, as my theory of offensive realism can, it might be able to say something useful about the future."[107]

Apart from offensive realism, two other international relations theories tended to support the view that the risk of conflict between China and other countries would grow in proportion to its power. The earliest of these was the "security dilemma", which Robert Jervis observed would arise because "many of the means by which a state tries to increase its security decreases the security of others". As each state reacted, this would lead to a spiralling competition.[108] A variation of the security dilemma was "power transition theory", which originated in the work of A. F. K. Organski.[109] He maintained that the international system was not anarchical but hierarchical. A stable international hierarchy was dominated by the most powerful nation, but could be destabilised when a revisionist challenger threatened to reach parity with the hegemon. These three theoretical approaches – offensive realism, the security dilemma, and power transition theory – all tended to accentuate fears about China's future international impact. Nevertheless they met with strong criticism from several quarters.

Many of the most plausible counter-arguments to offensive realism were marshalled by Zbigniew Brzezinski during an exchange with Mearsheimer. Noting that China's rise had been peaceful so far, he asserted that future conflict was "not inevitable or even likely" because its focus remained on economic development and on achieving great power status. It would therefore not want to challenge the US militarily. A focus on history and on "the iron laws of political theory" might point to "some inevitable collision or conflict", but he argued that there were other realities that would incline China to continue its cautious foreign policy. It gave priority to economic growth and knew that this could be disrupted by a confrontational policy, which might also threaten the CCP's hold on power. China's leadership appeared to him "much more flexible and sophisticated" than many previous great power aspirants. He also doubted it would have the military power to push the US out of Asia. Brzezinski therefore believed it would be possible for China to "avoid the negative consequences that often accompany the rise of new powers". He was "impressed by the power of theory" but thought that when events did not "fit the theory" the theory should be revised.[110]

Such a revision of theory took form in the work of Charles Glaser. In his book *Rational Theory of International Politics*, Glaser argued that "international anarchy did not generate a general tendency toward competitive international strategies" because in many circumstances "cooperation is a state's best option for achieving security". He distinguished between "security-seeking" states and "greedy" states. If a security-seeking state could feel secure within the international status quo, it would be more inclined to be cooperative than a greedy state that was "fundamentally dissatisfied" with the status quo. The security dilemma would be eliminated if both states were "sure that the opposing state is a security-seeker".[111] Glaser disputed realists' claims that China's "unique qualities, past behaviour, and economic trajectory" would drive it to become a destabilising force internationally. He believed that structural forces leading major powers into future conflict would be relatively weak and that disputes in Northeast Asia would be controllable due to the "security prevalent in the international system at large". In view of the "incentives for restraint and cooperation", conflict between the US and China was "not predetermined".[112]

Several other authors also expressed reservations about the application of international relations theory, and especially power transition theory, to predicting China's future path. For example, Steve Chan took a "contrarian" view to those who assumed the danger of war would increase as the "dissatisfied challenger" (China) caught up with the "existing hegemon" (the United States). He questioned why a "rising latecomer" would confront the pre-eminent power when differential growth rates alone would soon make it more powerful. He expected that rather than behave as a "cocky, rash upstart" China would probably try to avoid a "frontal collision".[113] Jeffrey Legro also questioned the persuasiveness of international relations theories. He agreed that there were "good reasons to believe" China's aims would grow as its power grew, but he countered that "power is not destiny". He pointed to a "large and very insightful" constructivist literature that showed the influence on foreign policy of "collective ideas" including beliefs, norms, discourses, and culture. He concluded that "the way China thinks about the world" would be an important influence.[114] Other researchers explored the potential for institutionalism and multilateralism to assist in "socialising" China and in paving the way for its cooperative participation in the international system.[115]

Experienced China watchers like Andrew Nathan and David Shambaugh also considered the prospects for strategic competition with China from the viewpoint of international relations theory, but chose to take an eclectic view. Nathan thought that realism – interpreted as emphasising the pursuit of national interests – often provided answers to the puzzles of China's foreign policy. However, he found that constructivism and institutionalism were also important influences.[116] Shambaugh judged that, whether from a realist, liberal or constructivist perspective, it was clear that international structures did much to shape US–China relations. Security structures tended to encourage competition, while economic and normative structures tended to constrain such competition. The high degree of interdependence between the two countries and the network of institutional links tended to reduce disruptions, but could also create frictions.[117] Shambaugh criticised an "increasing obsession" in the academic China community with social science theories that created "further impediments to understanding – more often blurring than clarifying the object of study".[118]

This discussion among Western experts raises the question of how international relations specialists within China itself have reacted to predications about its future behaviour that were developed by theorists outside China. Most were sceptical. They rebelled against the notion that China's future international interactions were somehow prescribed or predetermined by the findings of Western theory. One study involving both Chinese and US researchers considered that the behaviour of states was "not determined by a single variable" and that power transitions were also "not predetermined". The authors accepted that "persistent strategic competition" was likely, but war was "not the inevitable outcome". Each power transition was "a unique event" influenced by multiple factors that affected prospects for conflict or cooperation. In particular, the importance placed by rising powers on economic development could influence them to seek a cooperative international environment. They therefore felt optimistic that China and the US could avoid "a destructive power transition war".[119]

Other Chinese experts have supported these optimistic views. Yuan Peng claimed in 2012 that top leaders in both China and the United States shared the political will to build a relationship "that breaks with the old cycle of inevitable confrontation that accompanied the rise and fall of great powers in history", even though pessimism permeated "the media, society, and the military and even academic circles of the two sides". But he added that fewer and fewer international relations scholars were "playing the old refrain" of comparing China with Germany before World War One or Japan before World War Two. China was not a second Soviet Union and did not seek ideological expansion or to overthrow the existing international order, so Washington and Beijing would not "follow the same old disastrous road".[120] As Yuan conceded, however, not all Chinese scholars believed that China's transition would necessarily be peaceful. Yan Xuetong, for example, supported the concept of a harmonious international system as proposed in China's official foreign policy, but he did not believe this was achievable. He observed that "we realists don't believe the world can be harmonious" because "every power is greedy for leadership".[121]

Actors and perceptions

Theorists clearly were divided between realists, who believed material and structural factors were more important in determining China's international future, and institutionalists and constructivists, who believed the influence of organisations and ideas would be equally if not more important. Most China watchers, however, appeared willing to side with constructivists at least to the extent of acknowledging that the actions, strategies and perceptions of human players would certainly be a very important factor. It was also readily conceded, however, that the influence of those actors and their perceptions could be both positive and negative.

On the one hand, experts outside China acknowledged that both China's official foreign policy and many of its international relations experts supported a peaceful and cautious international stance. These experts rebelled against undue emphasis on the structural determinants of international politics. Rather, they emphasised interdependence, globalisation, and the need to tackle jointly a variety of new and pressing twenty-first century issues that would lead states to prefer cooperation over conflict. They eschewed the pursuit of hegemony and traditional great power competition. Wen Jiabao, during a visit to the US, once claimed that the argument that a rising power was "bound to seek hegemony" did not apply to China because this went against its "cultural tradition" and the will of its people.[122] On the other hand, concern was also expressed by outside experts about other competing influences on China's foreign policy, especially the upsurge of nationalism. US scholar Wang Zheng, for example, has argued that it is not possible to assess the future direction of China's international strategy without considering the critical influence of its historical memory of a Century of Humiliation. This memory had been cultivated and amplified in a "master historical narrative" promoted by a government-directed campaign of patriotic education, summarised by the slogan "Never Forget National Humiliation".[123]

Mutually negative perceptions in relations between China and the United States are especially problematic. Peter Gries noted that anti-American and anti-Chinese polemics were "pernicious" and could "easily spiral into mutual dehumanization and demonization, laying the foundation for violent conflict". In China a "conflictual view" seemed to have gained greater currency, with the US being viewed as "not just arrogant" but also actively seeking "to prevent China from prospering and gaining its rightful place at the top of the world system". Meanwhile in the US, according to Robert Ross, the policy debate was driven by "diametrically opposed understandings of Chinese intentions". Gries considered such polarisation to be "potentially disastrous", with US China bashers oblivious to their impact in China and Chinese nationalists responding with "equally virulent America bashing".[124] He took exception to realists like John Mearsheimer who dismissed the impact of perceptions on international politics, countering that the real "tragedy of world politics" was the failure of scholars and diplomats to recognise the importance of understanding how others viewed the world, "allowing misperceptions and conflicts to fester".[125]

Other experts were equally concerned about the influence of mutual perceptions on the prospects for competition or cooperation in US–China relations. David Shambaugh commented that neither the US nor China seemed "able to grasp complexity in the other", resulting in "mutual 'deafness' and unnecessary arrogance". He warned against demonising China.[126] Stanley Lubman found this issue to be especially problematic in the US Congress where various interests insisted on "strongly negative views of China that often caricature a complex society and foster unconstructive moralizing".[127] CIA analyst Josh Kerbel found that the policy and intelligence debate in the US over China was marred by "ingrained cognitive biases". He urged a departure from the "zero-sum perspectives, antagonism, hubris and stridency" that had been the "defining features" of the China debate in the US.[128]

Such concerns led some experts to give increased attention to the origins of strategic mistrust and how it might be alleviated. David Lampton pointed to several factors that were creating "a toxic mix that is corrosive to mutual trust and conducive to higher levels of future conflict".[129] A report co-authored by Kenneth Lieberthal and Wang Jisi included an exchange of views about the sources of distrust. On China's side, Wang said these included the view that the ultimate goal of the US was to "maintain its global hegemony" and to "constrain or even upset China's rise". China's leadership also felt threatened by the US "democracy promotion agenda". On the US side, Lieberthal believed that strategic distrust of China was "not the current dominant view" among decision-makers in Washington, but the US was concerned about PLA aspirations and suggestions that China might be "assuming a more hegemonic approach". Lieberthal and Wang agreed there would be "very high costs" if distrust continued to grow "at its current rapid pace".[130]

International futures

This chapter has explored a major theme in the expert debate about China's international future: namely, a persistent and all-consuming question as to whether its influence will be cooperative and peaceful or competitive and conflictual. It has concluded from this debate that this issue legitimately constitutes the dominant critical uncertainty in this domain. Not only is it the focus of so much of the expert debate, but its outcome is highly consequential for international cooperation and security. If the outcome should be that China's future international behaviour is indeed predominantly cooperative and peaceful – which is the stated objective of its declared foreign policy – then the interests of not only China but of all of its international partners will benefit greatly. If instead the outcome should be that this objective fails because China's actual agenda concentrates on the active and uncompromising fulfilment of broadly defined core national interests – and because other countries react to its growing power and influence with a similarly tough and adversarial posture – then the possible negative consequences in terms of military conflict, economic decline, and failure to address pressing global issues are almost too damaging to contemplate.

There is no doubt that perceptions of a competitive relationship between China and other major powers, especially the United States, have had a strong place in expert opinion in both China and other countries. Among experts outside China, the concept of a China Threat took root as its economy prospered during the 1990s, and blossomed following military tensions in the Taiwan Strait. However, other experts objected to "hype" about a belligerent China, disputed the China Threat concept, and argued that it would be too preoccupied internally to pose a serious threat externally. Many experts in China also objected to allegations that its rise would not be peaceful.

Subsequently, and especially following the GFC and the exacerbation of China's maritime disputes, concerns about an assertive China again became a major preoccupation among China experts in the West. Moreover, some of those who had earlier been inclined to downplay such concerns joined the growing negative consensus. Such negativity was fuelled by visible evidence of China's growing power, as well as by its apparently more nationalistic and less flexible posture regarding its core interests. One US expert concluded that future China–US relations would be likely to be based on "coopetition", combining both cooperation and competition but with competition in the ascendency. Other foreign experts suggested that successful management of US–China strategic competition might best be achieved by greater power-sharing, rather than military rebalancing.

China naturally had a different perspective on many of these issues and its official foreign policy continued to stress peaceful development. It experienced much difficulty, however, in convincing others of its benign intentions. Differences emerged among China's own international relations specialists about the extent to which it should adhere to its cautious and moderate posture or take advantage of the changing US–China power balance. Most of China's international relations experts endorsed a mainstream and officially endorsed position that emphasised China's interdependence with the US and the imperative for bilateral and international cooperation. However, even many of those who took this view nonetheless also appeared to accept the actual likelihood of increasingly competitive relations. Most of them also supported China's determination to promote its core interests, including its territorial claims.

In attempting to interpret and forecast China's strategic intentions, many Western experts went beyond empirical analysis of trends and drew upon international relations theories that they believed had predictive power. They pointed especially to the purported competitive consequences of the security dilemma concept and of power transition theory. Offensive realists argued that the drive for states to maximise their power would inevitably engender an intense security competition. Others disputed such deterministic theories, however, arguing that rival states might well conclude that their interests would be better served by cooperation. Institutionalists and constructivists saw further incentives for cooperation. Eclectics concluded that the major schools of international relations theory could all provide insights and that none should be relied upon exclusively. Chinese scholars also often objected to the predictive pretensions of Western international relations theory.

Despite their differences on other issues, there appeared to be a broad consensus among most China experts that the perceptions and strategies of key actors would exercise a powerful influence on the development of China's international relations. Many emphasised the destructive influence of strategic mistrust between China and the US, and were concerned about the impact of rising nationalism. One scholar, for example, maintained that China's future international relations could not be understood properly without consideration of the powerful influence of historical memory and patriotic education. Others drew attention to the damaging consequences of negative polemics and cognitive bias.

Contemporary expert opinion about China's future involvement within the Asia–Pacific region reflects considerable concern that a more difficult phase may have begun. Many actors in both the US and China appear disposed to discern malign intentions on the part of the other and to believe that the future relationship will be more competitive. Both sides are moving to strengthen their military capabilities. Some experts have voiced fears that a "slide" toward adversarial relationships may already have begun, and have lamented that there is insufficient awareness of the potentially catastrophic global consequences of such a trend. In this context realists could perhaps not be blamed if they were tempted to conclude that their theoretical perspective might be vindicated, and others might concede that some of China's actions are indeed fulfilling realist expectations. The growing trend toward confrontation between an increasingly powerful China and other countries over its maritime claims is one indication of the powerful logic behind theoretical concepts that emphasise the likely negative impacts of security dilemmas and power transitions.

It would be wrong, however, to assume that the majority of Western and Chinese experts, or their governments, have concluded that a future characterised by undiluted competition is inevitable. Quite the contrary, there continues to be recognition on both sides that, although competition will persist, its potentially disastrous consequences must and can be managed, and that despite the obstacles cooperation needs to be the principal policy objective. Many prominent experts within and outside China have supported the view that there is still scope for a predominantly cooperative future if there are appropriate attitudinal and policy adjustments on both sides, and that this is certainly preferable to the alternative. In the meantime, however, it is quite evident that whether China's future international influence will be mainly cooperative or competitive remains a vital and still unresolved critical uncertainty, ranking in importance and potentially negative consequences probably second only to developments in the environmental domain.

Notes

1 Ross H. Munro, "Awakening Dragon", *Policy Review* 62 (Fall 1992), 10–16.

2 Denny Roy, "Hegemon on the horizon?: China's threat to East Asian security", *International Security* 19, no. 1 (Summer 1994), 149–168. In a later article, Roy took a more impartial view: Denny Roy, "The 'China Threat' Issue: Major Arguments", *Asian Survey* 36, no. 8 (August 1996), 758–771.

3 Larry M. Wortzel, "China Pursues Traditional Great-Power Status", *Orbis* 38, no. 2 (Spring 1994), 157–175.
4 Goodman and Segal, *China Without Deng.*
5 Richard Bernstein and Ross H. Munro, *The Coming Conflict with China* (New York: Vintage, 1997), xviii–xix and 70–73, and Richard Bernstein and Ross H. Munro, "The Coming Conflict with America", *Foreign Affairs* 76, no. 2 (March/April 1997), 18–32.
6 Bernstein and Munro, *The Coming Conflict with China*, 19–21.
7 Samuel P. Huntington, "The Clash of Civilizations?", *Foreign Affairs* 72, no. 3 (Summer 1993), 45–47.
8 Samuel P. Huntington, *The Clash of Civilizations and the Remaking of World Order* (New York: Simon & Schuster, 1996), 229–232 and 312–316.
9 For example, Edward Timperlake and William C. Triplett II, *Red Dragon Rising: Communist China's Military Threat to America* (Washington, DC: Regnery, 1999); Bill Gertz, *The China Threat: How the People's Republic Targets America* (Washington, DC: Regnery, 2000); Steven W. Mosher, *Hegemon: China's Plan to Dominate Asia and the World* (San Francisco, CA: Encounter Books, 2000); and Jed Babbin and Edward Timperlake, *Showdown: Why China wants War with the United States* (Washington, DC: Regnery, 2006).
10 Constantine C. Menges, *China: The Gathering Threat* (Nashville, TN: Thomas Nelson, 2005), x–xvii and 367–417.
11 Charles Krauthammer, "Why we must contain China", *Time* (31 July 1995).
12 Robert Kaplan, "How We Would Fight China", *Atlantic Monthly* (June 2005).
13 Jonathan Pollack, "Chinese Military Power: What Vexes the United States and Why?", *Orbis* 51, no. 4 (2007), 635–650.
14 Andrew S. Erickson and Michael S. Chase, "An Undersea Deterrent?", *USNI Proceedings* 136, no. 6 (June 2009); Mark Stokes, *China's Evolving Conventional Strategic Strike Capability: The anti-ship ballistic missile challenge to US maritime operations in the Western Pacific and beyond* (Project 2049 Institute, 14 September 2009); Andrew S. Erickson and David D. Yang, "On the Verge of a Game-Changer", *USNI Proceedings* 135, no. 5 (May 2009); Andrew S. Erickson and David D. Yang, "Using the Land to Control the Sea?: Chinese Analysts Consider the Antiship Ballistic Missile", *Naval War College Review* 62, no. 4 (Autumn 2009), 53–86; and Toshi Yoshihara, "Chinese Missile Strategy and the US Naval Presence in Japan: The Operational View from Beijing", *Naval War College Review* 63, no. 3 (Summer 2010), 39–62.
15 Andrew S. Erickson, Abraham M. Denmark, and Gabriel Collins, "Beijing's 'Starter Carrier' and Future Steps: Alternatives and Implications", 65, no. 1 (Winter 2012), 50, and Nan Li and Christopher Weuve, "China's Aircraft Carrier Ambitions: An Update", *Naval War College Review* 63, no. 1 (Winter 2010), 28.
16 Dan Blumenthal, "Sino-US Competition and US Security: How Do We Assess the Military Balance?", *NBR Analysis* (December 2010), 4.
17 *Investigating the Chinese Threat, Part One: Military and Economic Aggression*, (United States Congress House Committee on Foreign Affairs, 28 March 2012). See also John J. Tkacik, *A Chinese Military Superpower?* (Heritage Foundation, 8 March 2007).
18 Andrew S. Erickson and Gabe Collins, "China Deploys World's First Long-Range, Land-Based 'Carrier Killer' ", *China SignPost* (26 December 2010).
19 Amy Chang, *Indigenous Weapons Development in China's Military Modernization* (US–China Economic and Security Review Commission, 5 April 2012).
20 Aaron L. Friedberg and Robert S. Ross, "Here Be Dragons: Is China a Military Threat?", *The National Interest* (September/October 2009).
21 James Dobbins, David Gompert, David Shlapak, and Andrew Scobell, *Conflict with China: Prospects, Consequences, and Strategies for Deterrence* (RAND Arroyo

Center, 2011), vii–viii. The argument that the US should cooperate with China from a position of military strength in the South China Sea was also emphasised in Patrick M. Cronin, *Cooperation from Strength: The United States, China and the South China Sea* (Center for a New American Security, January 2012), 5–6.
22 Andrew S. Erickson, *Through the Lens of Distance: Understanding and Responding to China's "Ripples of Capability"* (United States Naval War College, January 2012).
23 Andrew S. Erickson and Adam P. Liff, "A Player, but No Superpower", *Foreign Policy* (7 March 2013).
24 Nye, "China's Re-emergence and the Future of the Asia–Pacific", 68.
25 Keith Crane, Roger Cliff, Evan Medeiros, James Mulvenon, and William Overholt, *Modernising China's Military: Opportunities and Constraints* (RAND, 2005), xx and 101–110.
26 Richard Halloran and John J. Tkacik Jr, *Does China Pose a Military Threat?* (Council on Foreign Relations, 26 March 2007). See also John J. Tkacik Jnr, *Questioning the CIA's Claim of a Drop in China's Military Spending* (Heritage Foundation, 31 August 2007).
27 Dennis J. Blasko, "An Analysis of China's 2011 Defense Budget and Total Military Spending – The Great Unknown", *China Brief* 11, no. 4 (10 March 2011).
28 Andrew Davies and Andrew Rothe, *Theory and practice – A comparative look at China's new Defence White Paper* (Australian Strategic Policy Institute, 15 June 2011), 6 and 7. These estimates came from the Stockholm International Peace Research Institute.
29 Adam P. Liff, "Demystifying China's Defense Spending: Less Mysterious in the Aggregate", *China Quarterly* (March 2013), 1–26. See also Andrew S. Erickson and Adam P. Liff, "Understanding China's Defense Budget: What it Means, and Why it Matters", *Pacnet* 16 (9 March 2011).
30 *Defending Defense – China's Military Build-up: Implications for US Defense Spending* (A Joint Project of the American Enterprise Institute, the Heritage Foundation and the Foreign Policy Initiative, March 2011).
31 Wu Jiao, "China refutes military spending report", *China Daily* (10 July 2010). One survey found that over 80 percent of Chinese respondents supported higher military spending; see "Military Spending: Will ask, should tell", *Global Times* (5 March 2010).
32 Wu Jiao, "China refutes military spending report", *China Daily* (10 July 2010).
33 Carola Hoyos, "Chinese defence budget set to double by 2015", *Financial Times* (13 February 2012).
34 "The dating game", *The Economist*, (27 December 2011.)
35 Trefor Moss, "5 Things the Pentagon Isn't Telling US about the Chinese Military", *Foreign Policy* (23 May 2012).
36 Michael D. Swaine, *Interpreting China's Grand Strategy: Past, Present and Future* (RAND, 2000), ix–xii and 151–241.
37 Alastair Iain Johnston, "Is China a Status Quo Power?", *International Security* 27, no. 4 (Spring 2003), 5–55.
38 Minxin Pei, *Assertive Pragmatism: China's Economic Rise and its Impact on Chinese Foreign Policy* (Paris: IFRI, Fall 2006).
39 John Lee, *China's Insecurity and Search for Power* (St Leonards, NSW: Centre for Independent Studies, 13 November 2008).
40 Minxin Pei, "An Assertive China the 'New Normal'?" *The Diplomat* (24 November 2010).
41 Gordon Chang, "The Remilitarization of Beijing", *The Diplomat* (21 September 2010).
42 "Friend or foe?: A special report on China's place in the world", *The Economist* (4 December 2010).

43 David Pilling, "China will not be the world's deputy sheriff", *Financial Times* (27 January 2010).
44 David Shambaugh, "Coping with a Conflicted China", *The Washington Quarterly* 34, no. 1 (Winter 2011), 7–27.
45 Nick Bisley, "Biding and Hiding No Longer: A More Assertive China Rattles the Region", *Global Asia* 6, no. 4 (Winter 2011).
46 Aaron L. Friedberg, *A Contest for Supremacy: China, America, and the Struggle for Mastery in Asia* (New York: W.W. Norton, 2011), xiii–xiv and 1–4.
47 Ibid., 5–8, 33–57, 153–155, 215–244, and 264–284.
48 Aaron L. Friedberg, "Bucking Beijing", *Foreign Affairs* 91, no. 5 (September/October 2012), 48–58.
49 John Lee, "China's Rise and the Road to War", *Wall Street Journal* (5 August 2010).
50 David Shambaugh, ed. *Tangled Titans: The United States and China* (Lanham, MD: Rowman & Littlefield, 2013), xv–5 and 22. See also *Conceptualizing Future United States–China Relations*, (Washington, DC: Brookings Institution, 25 September 2012).
51 Jonathan Holslag, *Trapped Giant: China's Troubled Military Rise* (London, Institute of International and Strategic Studies, 2012), 134–139. See also Jonathan Holslag, "China's Vulnerability Trap", *Survival* 53, no. 2, (April-May 2011), 77–88.
52 Richard Bernstein, "China Basher Bashes Bashing", *New York Times* (29 June 1997).
53 David M. Lampton, "China: Think Again", *Foreign Policy*, no. 110 (Spring 1998).
54 Conference papers were collated in Ted Galen Carpenter and James A. Dorn, eds. *China's Future: Constructive Partner or Emerging Threat?* (Washington, DC: Cato Institute, 2000).
55 Michael D. Swaine, "Don't Demonize China: Rhetoric about its Military Might Doesn't Reflect Reality", *Washington Post*, (18 May 1997.) Similar arguments appeared in Avery Goldstein, "Great Expectations: Interpreting China's Arrival", *International Security* 22, no. 3 (Winter 1997), 36–73.
56 Nye, "China's Re-emergence and the Future of the Asia-Pacific".
57 Herbert S. Yee, "Into the New Century: Reconsidering the 'China Threat' Factor in Sino-US Relations", in *Return of the Dragon: US–China Relations in the 21st Century*, ed. Danny Paau and Herbert Yee (Frankfurt am Main: Peter Lang, 2005), 115–122. See also Herbert Yee and Ian Story, eds. The China Threat: Perceptions, Myths and Reality (London: Routledge Curzon, 2002), 21–38.
58 Emma V. Broomfield, "Perceptions of Danger: The China Threat Theory", *Journal of Contemporary China* 12, no. 35 (2003), 265–284.
59 David M. Lampton, "The Faces of Chinese Power", *Foreign Affairs* 86, no. 1 (January/February 2007), 115–127.
60 David M. Lampton, *The Three Faces of Chinese Power: Might, Money, and Minds* (Berkeley, CA: University of California Press, 2008), 2–7 and 252–260.
61 Harry Harding, "Think Again: China", *Foreign Policy* (March/April 2007).
62 Pei, "China's Not a Superpower". See also Shujie Yao, *Can China Really Become the Next Superpower?* (China Policy Institute, University of Nottingham, April 2007).
63 Andrew J. Nathan, "When China is no. 1", *What Matters* (McKinsey and Company, 26 February 2009). See also Joseph S. Nye, "American and Chinese Power after the Financial Crisis", *The Washington Quarterly* 33, no. 4 (2010), 143–153.
64 Yong Deng, *China's Struggle for Status: The Realignment of International Relations* (New York: Cambridge University Press, 2008), 22–24 and 97–127.
65 Michael D. Swaine, "Perceptions of an Assertive China", *China Leadership Monitor* 32 (Spring 2010). Other articles discussing China's assertiveness were published in subsequent issues of *China Leadership Monitor* in 2011 and 2012.

66 Thomas J. Christensen, "The Advantages of an Assertive China", *Foreign Affairs* 90, no. 2 (March/April 2011), 54–67.
67 Andrew J. Nathan, "What China Wants", *Foreign Affairs* 90, no. 4 (July/August 2011), 153–158.
68 Andrew Scobell and Andrew J. Nathan, "China's Overstretched Military", *The Washington Quarterly* 35, no. 4 (2012), 135–148. See also Andrew J. Nathan and Andrew Scobell, *China's Search for Security* (New York: Columbia University Press, 2012), xi–xiii and 345–359.
69 Lyle J. Goldstein, "Resetting the US–China Security Relationship", *Survival* 53, no. 2 (2011), 89–92 and 111.
70 Michael D. Swaine, *America's Challenge: Engaging a Rising China in the Twenty-First Century* (Washington, DC: Carnegie Endowment for International Peace, 2011), 15–19, 34–51, 147–182 and 337–380.
71 Henry A. Kissinger, "China: Containment Won't Work", *Washington Post* (13 June 2005).
72 Henry A. Kissinger, "The Future of US–Chinese Relations", *Foreign Affairs* 91, no. 2 (March/April 2012), 44–55.
73 Henry A. Kissinger, *On China* (London: Penguin, 2012), 520–530.
74 Joseph S. Nye, "A New Great Power Relationship?", *China–US Focus* (26 February 2013). Other experts who were optimistic about a cooperative future between China and other countries included US intelligence analyst Eric Anderson and British scholar Barry Buzan. See Anderson, *China Restored: The Middle Kingdom Looks to 2020 and Beyond*, xi-xxiii and 249–250; Barry Buzan, The Inaugural Kenneth N. Waltz Annual Lecture, "A World Without Superpowers: Decentred Globalism", *International Relations* 25, no. 3 (2011), 3–25.
75 Hugh White, "Australian Defence Policy and the Possibility of War", *Australian Journal of International Affairs* 56, no. 2 (2002), 260, and Hugh White, "The Limits to Optimism: Australia and the Rise of China", *Australian Journal of International Affairs* 59, no. 4 (December 2005), 469–480.
76 Hugh White, "The Geo-strategic Implications of China's Growth", in *China's New Place in a World in Crisis: Economic, Geopolitical and Environmental Dimensions*, ed. Ross Garnaut, Ligang Song, and Wing Thye Woo (Canberra: Australian National University E Press, 2009), 89–102; White, "Power Shift: Australia's Future Between Washington and Beijing".
77 Hugh White, *The China Choice: Why America should share power* (Collingwood, VIC: Black Inc., 2012), 1–10.
78 Kevin Rudd, "Beyond the Pivot", *Foreign Affairs* 92, no. 2 (March/April 2013), 9–15.
79 Zheng Bijian, "China's 'Peaceful Rise' to Great Power Status", *Foreign Affairs* 84, no. 5 (September/October 2005).
80 Wang Jisi, "What China Needs in the World and from the United States", in *China's March on the 21st Century*, ed. Kurt M. Campbell and Willow Darsie (The Aspen Institute, 2007), 85–92.
81 Wang Jisi, "China's Search for Stability with America", *Foreign Affairs* 84, no. 5 (September/October 2005), 39–48.
82 Wang Jisi, "America in Asia: How much does China care?", *Global Asia* 2, no. 2 (Fall 2007).
83 Jia Qingguo, "Peaceful development: China's policy of reassurance", *Australian Journal of International Affairs* 59, no. 4 (December 2005), 493–507.
84 Zhu Feng, "China's Rise will be Peaceful: How Unipolarity Matters", in *China's Ascent: Power, Security and the Future of International Politics*, ed. Robert S. Ross and Zhu Feng (Ithaca, NY: Cornell University Press, 2008), 34–36 and 52–54.
85 Andrew J. Nathan and Andrew Scobell, "How China Sees America", *Foreign Affairs* 91, no. 5 (September/October 2012), 32–47. For further analysis see Roger Irvine,

"Primacy and Responsibility: China's Perception of Its International Future", *China Security* 6, no. 3 (2010), 23–42.
86 Wang Jisi, "China's Search for a Grand Strategy", *Foreign Affairs* 90, no. 2 (March/April 2011), 68–79.
87 Wang Jisi, "China's grim international environment", in Mark Leonard, ed. *China 3.0* (London: European Council on Foreign Relations, November 2012), 118–124.
88 Zhu Feng, "No One Wants a Clash", *New York Times* (3 May 2012).
89 Lin Limin, "Future International Trends as Predicted from the International Situation of 2011", *Contemporary International Relations* (January/February 2012), 1–22.
90 Yan Xuetong, "The weakening of the unipolar configuration", in Leonard, *China 3.0*, 112–117.
91 B. Creutzfeldt, "Yan Xuetong on Chinese Realism, The Tsinghua School of International Relations, and the Impossibility of Harmony", *Theory Talks* 51 (2012).
92 Yan Xuetong, "The Instability of China–US Relations", *Chinese Journal of International Politics* 3 (2010), 263–292.
93 Yan Xuetong and Qi Haixia, "Football Game Rather Than Boxing Match: China–US Intensifying Rivalry Does not Amount to Cold War", *Chinese Journal of International Politics* 5 (2012), 105–127.
94 As reported in *Global Times* "Bold predictions about China's future discern inertia of history" (8 August 2013).
95 Yan Xuetong, "New Values for New International Norms", *China International Studies* 38, no. 1 (January/February 2013).
96 Wu Xinbo, "Chinese Visions of Future US–China Relations", in Shambaugh, *Tangled Titans*: *The United States and China*, 371–388.
97 Wu Xinbo, "Not Backing Down: China Responds to the US Rebalance in Asia", *Global Asia* 7, no. 4 (Winter 2012).
98 Lin Limin, "Future International Trends as Predicted from the International Situation of 2011".
99 Lin Limin, "Breaking the Territorial Impasse: A Strategic Priority for China", *Contemporary International Relations* (September/October 2012), 98–102.
100 Lin Limin, "Sino-US Relations 'Paradigms' and China's US Strategy", *Contemporary International Relations* (May/June 2012), 1–19.
101 Yang Yi, "The China–US Relationship Does Not Need Trouble-Makers", *China–US Focus* (13 January 2013).
102 John Mearsheimer, *The Tragedy of Great Power Politics* (New York: W. W. Norton, 2001).
103 Ibid., 396–402.
104 John Mearsheimer, "China's Unpeaceful Rise", *Current History* (April 2006), 160–162.
105 John Mearsheimer, "The Gathering Storm: China's Challenge to US Power in Asia", *Chinese Journal of International Politics* 3 (2010), 381–396.
106 John Mearsheimer, "Can China, US avoid tragedy of great power politics", *Global Times* (27 May 2012). In this interview Mearsheimer estimated that 50 percent of experts within the US government and 75 percent of US academics agreed with his views.
107 Robert D. Kaplan, "Why John J. Mearsheimer is Right (About Some Things)", *Atlantic Monthly* (January/February 2012).
108 Robert Jervis, "Cooperation Under the Security Dilemma", *World Politics* 30, no. 2 (January 1978), 169.
109 A. F. K. Organski, *World Politics* (New York: Knopf, 1958).
110 Zbigniew Brzezinski, "Clash of the Titans", *Foreign Policy* (January/February 2005).
111 Charles L. Glaser, *Rational Theory of International Politics*: *The Logic of Competition and Cooperation* (Princeton, NJ: Princeton University Press, 2010), 9–12.

112 Charles Glaser, "Will China's Rise Lead to War?", *Foreign Affairs* 90, no. 2 (March/April 2011), 80–91.
113 Steve Chan, *China, the US, and the Power-Transition Theory: A Critique* (London: Routledge, 2007), ix–xii and 1–10.
114 Jeffrey W. Legro, "What China Will Want: The Future Intentions of a Rising Power", *Perspectives on Politics* 5, no. 3 (September 2007), 515–534.
115 Avery Goldstein, "Power Transitions, Institutions, and China's Rise in East Asia: Theoretical Expectations and Evidence", *Journal of Strategic Studies* 30, no. 4–5 (August–October 2007), 639–646; Guoguang Wu and Helen Lansdowne, eds. *China Turns to Multilateralism: Foreign Policy and Regional Security* (London: Routledge, 2008); and Pauline Kerr, Stuart Harris, and Qin Yaqing, eds. *China's "New" Diplomacy: Tactical or Fundamental Change?* (New York: Palgrave Macmillan, 2008).
116 Nathan and Scobell, *China's Search for Security*, xvi. This is reminiscent of Katzenstein and Okawara's advocacy of "analytical eclecticism" in Peter J. Katzenstein and Noburo Okawara, "Japan, Asian–Pacific Security, and the Case for Analytical Eclecticism".
117 Shambaugh, *Tangled Titans*, 20.
118 David Shambaugh, *China Goes Global: The Partial Power* (New York: Oxford University Press, 2013), ix–x.
119 Robert S. Ross and Zhu Feng, eds. *China's Ascent: Power, Security, and the Future of International Politics* (Ithaca, NY: Cornell University Press, 2008), 1–8 and 305–315.
120 Yuan Peng, "Some Strategic Thoughts on New Type China–US Ties", *Contemporary International Relations* (July/August 2012).
121 Creutzfeldt, "Yan Xuetong on Chinese Realism".
122 Wang Zheng, *Never Forget National Humiliation: Historical Memory in Chinese Politics and Foreign Relations* (New York: Columbia University Press, 2012), 239.
123 Ibid., xiii–2 and 237–242.
124 Peter Hays Gries, *China's New Nationalism: Pride, Politics, and Diplomacy* (Berkeley, CA: University of California Press, 2004), 1–12 and 135–150.
125 Peter Hays Gries, "Problems of Misperception in US–China Relations", *Orbis* 53, no. 2 (Spring 2009), 220–232.
126 David Shambaugh, "Imagining Demons: The rise of negative imagery in US–China relations", *Journal of Contemporary China* 12, no. 35 (2003), 235–237.
127 Stanley Lubman, "The Dragon as Demon: Images of China on Capital Hill", *Journal of Contemporary China* 13, no. 40 (2004), 541–565.
128 Josh Kerbel, "Thinking Straight: Cognitive Bias in the US Debate About China", 27–35.
129 David M. Lampton, "Power Constrained: Sources of Mutual Strategic Suspicion in US–China Relations", *NBR Analysis* (June 2010).
130 Kenneth Lieberthal and Wang Jisi, *Addressing US–China Strategic Distrust* (Washington, DC: Brookings Institution, March 2012). See also Douglas G. Spelman, ed. *The United States and China: Mutual Public Perceptions* (Washington, DC: Woodrow Wilson International Center for Scholars, 2011), Preface and Parts 1 and 2.

Bibliography

Bernstein, Richard and Ross H. Munro. *The Coming Conflict with China*. New York: Vintage, 1997.

Carpenter, Ted Galen and James A. Dorn, eds. *China's Future: Constructive Partner or Emerging Threat?* Washington, DC: Cato Institute, 2000.

Chan, Steve. *China, the US, and the Power-Transition Theory: A Critique*. London: Routledge, 2007.

Deng, Yong. *China's Struggle for Status: The Realignment of International Relations*. New York: Cambridge University Press, 2008.

Fravel, Taylor M. *Strong Borders Secure Nation: Cooperation and Conflict in China's Territorial Disputes*. Princeton, NJ: Princeton University Press, 2008.

Friedberg, Aaron L. *A Contest for Supremacy: China, America, and the Struggle for Mastery in Asia*. New York: W. W. Norton, 2011.

Gertz, Bill. *The China Threat: How the People's Republic Targets America*. Washington, DC: Regnery, 2000.

Glaser, Charles L. *Rational Theory of International Politics: The Logic of Competition and Cooperation*. Princeton, NJ: Princeton University Press, 2010.

Goldstein, Avery. *Rising to the Challenge: China's Grand Strategy and International Security*. Stanford, CA: Stanford University Press, 2005.

Gries, Peter Hays. *China's New Nationalism: Pride, Politics, and Diplomacy*. Berkeley, CA: University of California Press, 2004.

Holslag, Jonathan. *Trapped Giant: China's Troubled Military Rise*. London, Institute of International Strategic Studies, 2012.

Huntington, Samuel P. *The Clash of Civilizations and the Remaking of World Order*. New York: Simon & Schuster, 1996.

Kerr, Pauline, Stuart Harris and Yaqing Qin, eds. *China's "New" Diplomacy: Tactical or Fundamental Change?* New York: Palgrave Macmillan, 2008.

Lampton, David M. *The Three Faces of Chinese Power: Might, Money, and Minds*. Berkeley, CA: University of California Press, 2008.

Lee, John. *China's Insecurity and Search for Power*. St Leonards, NSW: Centre for Independent Studies, 13 November 2008.

Leonard, Mark, ed. *China 3.0*. London: European Council on Foreign Relations, November 2012.

Lieberthal, Kenneth and Wang Jisi. *Addressing US–China Strategic Distrust*. Washington, DC: Brookings Institution, March 2012.

Mearsheimer, John J. *The Tragedy of Great Power Politics*. New York: W. W. Norton, 2001.

Menges, Constantine C. *China: The Gathering Threat*. Nashville, TN: Thomas Nelson, 2005.

Nathan, Andrew J. and Andrew Scobell. *China's Search for Security*. New York: Columbia University Press, 2012.

Pei, Minxin. *Assertive Pragmatism: China's Economic Rise and Its Impact on Chinese Foreign Policy*. Paris: IFRI, Fall 2006.

Ross, Robert S. and Zhu Feng, eds. *China's Ascent: Power, Security, and the Future of International Politics*. Ithaca, NY: Cornell University Press, 2008.

Shambaugh, David. *China Goes Global: The Partial Power*. New York: Oxford University Press, 2013.

Shambaugh, David, ed. *Tangled Titans: The United States and China*. Lanham, MD: Rowman & Littlefield, 2013.

Swaine, Michael D. *America's Challenge: Engaging a Rising China in the Twenty-First Century*. Washington, DC: Carnegie Endowment for International Peace, 2011.

Swaine, Michael D. *Interpreting China's Grand Strategy: Past, Present and Future*. RAND, 2000.

Wang Jisi. "What China Needs in the World and from the United States". In *China's March on the 21st Century*, edited by Kurt M. Campbell and Willow Darsie: The Aspen Institute, 2007.

Wang, Zheng. *Never Forget National Humiliation: Historical Memory in Chinese Politics and Foreign Relations*. New York: Columbia University Press, 2012.

White, Hugh. *The China Choice: Why America Should Share Power*. Collingwood, VIC: Black Inc., 2012.

Yee, Herbert and Ian Story, eds. *The China Threat: Perceptions, Myths and Reality*. London: RoutledgeCurzon, 2002.

8 Embracing uncertainty

This book has undertaken a systematic evaluation of expert forecasts of China's future, and of the methods employed, in the hope that this will suggest improved approaches for this challenging but important endeavour. Previous chapters have presented a representative sampling of the spectrum of expert opinion about the major forecasting issues, identifying the principal arguments and their main proponents. From this perspective, two important conclusions can be drawn. The first of these is that a high level of uncertainty inevitably will continue to surround most of the key issues confronting China. This uncertainty compromises the effectiveness of all of the most commonly used forecasting methods. It also highlights the absurdity of expert judgements that fail to acknowledge this uncertainty and to consider the merits of competing arguments.

The second conclusion is that the crucial challenge facing China experts is to find more effective means to respond to this prevailing uncertainty. Attention has been drawn not only to the many shortcomings of most of the traditional methods but also to possible ways in which they can be ameliorated. But the principal recommendation is that there is likely to be considerable advantage in an alternative forecasting approach that places uncertainty at the forefront, especially by focussing attention on the investigation of a structured framework of critical uncertainties.

Forecasting methods and limitations

All methods for forecasting the futures of nations are seriously handicapped by the inescapable fact that these futures are necessarily uncertain. Two of the methods that this study has investigated are central to understanding the sources of this unpredictability. First, the influence of human actors will continue to be one of the most powerful contributors to uncertainty. China's future direction seems likely to be determined more than almost anything else by the strategies of key actors, both within China and in the countries with which it interacts. Second, unpredictable surprise factors will also inevitably continue to have a major impact, and may greatly modify the scope for these actors to shape events. Due recognition therefore needs to be given in all forecasts to the often decisive influence of both human actions and surprise developments.

Of the remaining forecasting methods, trend analysis is an obvious and fundamental technique, although there has been much discussion about the perceived inadequacy of forecasts that tend to assume the linear continuation of historical trends. By far the most controversial method, however, has been the application of expert opinion to this analysis of trends. Individual experts frequently overestimate the level of certainty that is achievable and underestimate the scope for alternative outcomes. The chief difficulty in relying on such expert judgement is readily apparent in earlier chapters; namely, that very different and even diametrically opposed opinions are often expressed by rival experts on the same subject. Even more strikingly, a minority of protagonists at both ends of the spectrum often considered their extreme and contested opinions about future outcomes to be both certain and inevitable. The mutual incompatibility of such opinions has seriously degraded the value and credibility of the collective endeavours of China experts. They produced forecasting results that were at best mixed and at worst dramatically at variance with actual outcomes.

Many experts also relied heavily – and often excessively – on predictive theories and predictive models to assist them to discern the future and attempted to use these theories and models to bolster the credibility of their forecasts. Predictive *theories* were most popular in the political, economic, and international relations domains. In virtually all cases, however, the deterministic pretensions and credibility of these theories were challenged by other experts. The latter conceded that such theories could sometimes offer useful or even powerful insights but dismissed the fanciful idea that they could be relied upon to predict the future confidently.

Predictive *models* were usually employed to forecast aspects of China's future quantitatively, although they proved to be equally as fallible as predictive theory. Statistical models were used to forecast economic growth and the comparative size of national economies over time, as well as to project energy consumption and carbon emissions. There were many spectacular examples of forecasting failures in these areas – especially the underestimation of China's GDP growth and of its energy and emissions trends. Few would suggest, however, that these predictive models should be abandoned entirely. When used intelligently and iteratively, and when alternative modelling assumptions were incorporated, they often provided indispensable guidance for policy- and decision-makers. Nevertheless, there is certainly much room for improvement in modelling techniques – including by considering a wider range of variables, accepting the fallibility of multiple assumptions, and acknowledging a wider margin for error.

Scenario construction also offers a potentially promising method to address the shortcomings of other approaches. Its particular merit is that it explicitly accepts the existence of high levels of uncertainty and encourages the consideration of a variety of possible alternative outcomes. It seeks to alert decision-makers to the potential for them to influence those outcomes through their future policies and actions. Scenario construction was often used to good effect by China experts, although it is debatable whether it achieved the level of impact on decision-making that some of its proponents envisaged. Scenario exercises were often handicapped by an understandable reluctance to assign probabilities to the diverse potential outcomes that were identified. Others were limited by a predisposition toward negative outcomes. However, when used skilfully by knowledgeable experts, scenario constructions were capable of producing useful results that stood the test of time – including in some instances where experts were brave enough to attempt to assign probabilities.

In summary, several viable forecasting methods were available to China experts. Their validity has been demonstrated, at least in principle. In many instances they produced valuable and influential assessments. None are likely to be abandoned in the future. Some of these methods are essential and even unavoidable. In combination, and when used flexibly and selectively over time, they assisted China experts to anticipate likely or possible futures and prompted others to plan and act accordingly. However, in practice they also exhibited many shortcomings. Experts often placed greater confidence in their efficacy than was justified. All of the methods proved to be fallible. These disappointing results are not surprising given the inherent impossibility of accurately predicting the future, and especially the future of such a dynamic country as China.

Experts and uncertainty

Clearly, *methods* for forecasting China's future have been problematical, and the *results* of their actual employment by experts to forecast specific major issues have been even more so. It is difficult to generalise about these forecasts without appearing unduly critical of the dedicated efforts of the majority of China experts, or failing to do justice to the complexity of their reasoning. However, it is nonetheless obvious that their achievements fell well short of their aspirations. Failings ranged from simplistic linear extrapolation to insufficient anticipation of non-linearities, from insufficient depth of empirical research to either conscious or unconscious bias, from excessive certitude about ultimate outcomes to sometimes overcautious even-handedness, and from unwillingness to adjust judgements in the face of persistent contrary information to an apparent aversion to considering the merits of competing arguments.

Other failings are more ambiguous. Given that uncertainty posed such intimidating difficulties for China watchers, it is perhaps not surprising that their forecasts were often imprecise and therefore difficult to evaluate. Timeframes were often unclear or even unspecified, with frequent resort to dubious expressions such as "for the foreseeable future". Many forecasts were open-ended. It could be difficult to say they were incorrect even if they seemed unlikely in the present, because if future circumstances changed they could be vindicated. Other forecasts were sufficiently imprecise that they could be interpreted in several different ways and their accuracy could therefore easily be disputed after the event. Examples include claims that aspects of China's development are trapped, or that the CCP will survive because of its adaptability, or that China will become dominant in its region, or its growing power will make it more demanding. Such outcomes tend to be relative rather than absolute, subjective rather than objective.

The most problematic and troubling aspect of forecasts by China experts collectively, however, has been demonstrated repeatedly in this study. Namely, not only was there a continuing polarisation of contradictory pessimistic and optimistic forecasts about China's overall national prospects, but experts also tended to produce opposing and often equally persuasive opinions about almost all of the specific major issues within each domain. Just as those who held gloomy or confident views about China's overall national prospects could not both be right, nor could they when expressing firm but competing views on specific domain-level issues – such as the prospects for authoritarianism or political liberalisation, for national unity or social instability, for economic growth or decline, for environmental progress or failure, and for peace or conflict internationally. China experts may be able to rationalise or tolerate such ambiguities, but, for policy-makers and for the many others who wish simply to be better informed, it is very difficult to assess which of the competing expert views are most worthy of consideration.

It would seem logical, given the many obvious imponderables, that experts would see the need to consider future possibilities more even-handedly and to adopt a more balanced and nuanced approach that considered competing viewpoints. Instances of this were comparatively uncommon, however. More frequently, they seemed to prefer to assemble an apparent preponderance of evidence and argument that tended to merely confirm their particular proclivities or biases. Individual experts appeared resistant to fully investigating and presenting both sides of an issue, before rushing precipitately to sanguine or alarmist judgements. Some seemed almost to relish gloomy outcomes and were quick to disparage those who saw scope for positive scenarios. These behaviours are typical of the traits of the opinionated and defensive hedgehogs so aptly described by Philip Tetlock. Such experts often focused their energies on "one big thing" on which they built their reputations. They refused to revise their opinions when challenged and declined to admit their forecasting failures.

This was by no means the whole story, however. Many other China experts – including probably a majority of the most experienced and most professional China watchers – displayed a much greater respect for uncertainty. They were willing to consider alternative possibilities and were prepared to adjust or qualify their views over time as new evidence emerged. For many such experts, the notion of paying greater attention to inherent uncertainties was very much a matter of preaching to the converted. They often expressed considerable frustration about the sensational analyses of some of their colleagues who they felt tended to discredit their field. Their outlook was much more akin to Tetlock's foxes who were sceptical about grand theories, more reserved in their forecasts, and willing to adjust their views as circumstances changed.

As indicated in Chapter 3, many of the most discerning China watchers favoured a view of its future that emphasised uncertainty rather than a distinctly negative or positive view. They often drew similar conclusions about the way experts should think and conduct themselves in these circumstances. This led them sometimes to make insightful attempts to summarise this distilled wisdom.

For example, Josh Kerbel approved of Jonathan Spence's comment that "the more blurred and multifaceted our perceptions of China become, the closer we may be to that most elusive thing: the truth".[1] Cheng Li insisted that, to produce rigorous and insightful assessments, students of China had to acquire "the intellectual ability to live with complexity, tolerate ambiguity, and expect uncertainty".[2] Orville Schell admitted that he did not know where things were going in China because they were "always going in opposite directions at the same time". He thought there was "no understanding the place" without being able to "embrace such contradictions in your head at the same time". In this he was supported by Beijing-based blogger Bill Bishop, who advised that the "most important skill" for understanding China was to master "cognitive dissonance". Such thinking was also perhaps what led Michel Oksenberg to observe that China was "never as good as it appears in its best moments, and never as bad as it appears in its worst moments".[3] Jonathan Fenby attributed to sinologist Richard Baum the view that in China it was perfectly possible to "come up with an idea in the morning and find all the evidence you need to take the opposite position by the afternoon".[4]

Critical uncertainties

The emphasis that many of the world's most respected China watchers placed on the degree of uncertainty surrounding most important aspects of China's future is striking. This suggested to me that the most constructive starting point for consideration of the domain-level expert debate on China's political, economic, environmental, and international issues was to treat uncertainty as the central issue and the main analytical focus. In particular, it seemed evident that the best approach was to concentrate this analysis on presenting the contrasting opinions of a representative selection of prominent experts on each of the major issues. The frequency with which I encountered passing references to the term "critical uncertainties" prompted me to adopt this terminology. But it also seemed highly advantageous to extend the scope of this promising concept by concentrating especially on those critical uncertainties that are the most strategic and consequential for China. This approach focuses attention squarely on the decisive influence that the most potent uncertainties will have on future outcomes. The purpose of articulating these critical uncertainties is to explore systematically the possible answers to essentially open questions about those issues that will have very high impact both for China and for the world. They can probably best be presented and researched as a hierarchy of domain-specific interrogative questions. These questions – revised as appropriate over time – serve both to identify the major issues under debate and to stimulate balanced and iterative consideration of opposing possibilities.

This approach was amply vindicated by the findings of Chapters 4 to 7. Side-by-side comparison of the judgements of the most influential experts on particular critical uncertainties in each of the four domains proved to be a most effective way to expose and test the credibility of alternative outcomes. Especially notable was the conspicuous phenomenon that, rather than one expert viewpoint or another being dominant in the literature, in the case of virtually all of the critical uncertainties examined there was a respectable body of analysis supporting each of the alternative possibilities. With surprising regularity it emerged that there were strong forces arguing the case for both positive and negative outcomes. This often exposed the folly of experts who regularly claimed sufficient perspicacity to be able to see above the conflicting evidence and to confidently predict one outcome over another. Such over-confidence frequently appeared to be presumptuous and unwarranted.

A further major advantage of an approach that accepts the prevalence of uncertainty is that it empowers a whole new perspective. It emphasises the reality that, for the majority of issues, both positive and negative outcomes are possible. Rather than creating an intolerable dilemma, this perspective has the potential to persuade analysts and policy-makers to see uncertainty as an opportunity. It highlights the practicality and achievability of positive and desired outcomes, rather than merely negative and defeatist outcomes. It highlights the realistic possibility that the odds in favour of positive outcomes may be increased if appropriate policy responses are adopted – within China, by other countries unilaterally, and through intensified cooperation between China and other countries collectively.

The analytical framework applied in Chapters 4 to 7 had special value in that it directed attention not only to well-known areas of divergent expert opinion but also to important and less-publicised forecasting results that have received comparatively little attention. In the political domain, it became clear from the contending forecasts that there is no real consensus of expert opinion about the longer-term prospects for China's currently well-entrenched authoritarian system. Although in the past the CCP has proved to be sufficiently skilful at adapting the political system to the demands of its citizens, many experts raise important questions about its ability to continue to adapt successfully as popular expectations evolve and escalate. Many experts consider that China's future political direction will be decided primarily by internal forces, and that it is unlikely to duplicate Western-style democracy. But whether continued insistence on its monopoly of political power will permit the CCP to maintain a sufficient level of legitimacy in the face of endemic challenges (such as pervasive corruption and inequality) is considered by many to be doubtful, even though most would concede that major political instability is unlikely at present given the CCP's vigilance and counter-measures. Most experts consider that some form of authoritarian regime will probably continue and that political liberalisation will be constrained, but others see potential for greater liberalisation and even a long-term prospect for full-blown democracy. These contradictory findings undermine the confidence that can be placed in analyses that imply either that the CCP's future is secure or that its eventual demise is likely.

In the economic domain, uncertainty was also revealed to be uppermost. It is true that China's capacity for economic management has been repeatedly underestimated, but whether the regime will continue to perform at the level required is still a major question. Will the leadership succeed in overcoming powerful vested interests in the Party-state as well as in big business, and transition to a new economic model that will deliver somewhat slower but higher-quality growth? Most experts expect at least mixed success, but a minority foresee a sharp decline. There also remains a substantial level of scepticism among outside experts about when – or even whether – China's GDP will overtake the US, despite that fact that even with slower growth it appears to be rapidly closing the gap. Many economic experts are also reluctant to accept the significant arguments in favour of using purchasing power parity rather than fluctuating market exchange rates to compare China's economy with the US, which would imply that the gap would close even more quickly. But those experts, including in China, who point out that this would still leave its per capita income far below the US, highlight a significant limitation on China's future capability.

Expert forecasts about prospects in the remaining two environmental and international domains have often been notably pessimistic. Outside China, popular expectations of its future environmental performance are typically low. Its past performance is considered to have been very poor, its current environmental impact is considered to be alarming, and its ability to turn that situation around is also regarded very sceptically. This popular assessment is shared – and for good reason – by many experts, who are aware of the reality that China is now, by a considerable margin, the largest contributor to global carbon emissions as well as the biggest contributor to the growth of those emissions. There has been less appreciation, however, of China's impressive and determined efforts to constrain its emissions and even possibly to cause them to peak within the next one to two decades. Such research is mainly a preoccupation of quantitative forecasters who have devoted their attention to this specialised subject. Popular and even expert perceptions of China's environmental measures are handicapped by considerable ignorance and inattention. It seems compelling that significantly more attention needs to be paid to further evaluation of the crucial question as to whether China's environmental counter-measures will have sufficient impact to rein in its still rapidly expanding emissions. In addition, it seems obvious – in the self-interest of all parties – that China's positive steps to date should be given more encouragement and support by the international community.

In the international domain, current trends are also of great concern. The overwhelming interests of China, the United States and other countries in cooperative future relationships remain at risk of being undermined by a slide toward distrust and rivalry, and even by the possible development of an intense strategic competition with potentially disastrous consequences. Powerful influences are at play that are exacerbated – on all sides – by historical narratives, nationalism, exceptionalism, territorial disputes, military expansion, and perceptions of relative power. This trend may be temporary, but even if it can be reduced there is likely still to be considerable room for future misunderstandings, miscalculations, and military confrontations that will need skilful management and a new emphasis on improved communications, power-sharing, and alleviation of strategic distrust. The expert debate about these international uncertainties is intense and prolific, both inside and outside China. Many recognise that both competition and cooperation are likely, and that a delicate balance will be required. Others are distrustful and pessimistic. They are concerned that conflict is a real possibility, whether by accident or as a deliberate instrument of policy. At the other end of the spectrum, some experts may be unduly confident about the prospects for cooperation, even though that path seems by far the most beneficial for the national interests of all concerned. In any case it is clear that uncertainty – and disagreement among experts – is equally as characteristic of the international domain as it is of other domains.

To summarise, the diversity of expert opinion revealed in Chapters 4 to 7 confirms the major benefits likely to be expected from an analytical framework based on the balanced assessment of critical uncertainties. This would mean relying less exclusively on individual expert opinion and much more on consideration of the diverse opinions of different experts. This framework for analysis accepts, realistically and constructively, that on most major issues future trends remain undetermined. Most of the issues still hang in the balance and can be influenced in one direction or another, especially by alternative strategies and human interventions in both China and other countries.

These conclusions lead to another, namely that a more *collective* approach to expert analysis of critical uncertainties is warranted as a supplement to individual endeavour. Efforts to bring experts together to synthesise different viewpoints, while still respecting individual opinions, would permit a more integrated approach for the evaluation of disparate expert forecasts. This approach could also usefully include periodic Delphi-like exercises to attempt to identify areas of expert consensus or dissensus on key issues. A more *institutionalised* approach in support of such collective endeavours also seems likely be beneficial. It is possible to envisage, for example, the establishment of new research centres (or virtual centres) dedicated to iterative and structured research about China's future, based on a developing framework of identified critical uncertainties. There are currently few examples of academic or other public institutions that devote their attention primary to such macro-level research on possible national futures – either for China or for other key countries.

Without claiming that the approach recommended here in favour of greater concentration on the investigation of critical uncertainties is the *only* right and proper way forward for analysis of China's future, I believe that this study has shown convincingly that this approach nonetheless offers a positive and effective way forward. At the least, the findings presented here lead overwhelmingly to the conclusion that an approach that willingly embraces the essential uncertainty of critical future outcomes in China is bound to be greatly superior to past reliance on a conspicuous diversity of individual expert opinion. Such a new approach could have more in common with the preference of many Chinese intellectuals for a dialectical perspective in which trends are viewed as being driven by a continuing competition between contradictory forces, compared to typically Western approaches that seek to identify a single or dominant direction.

An objection may be raised that such seemingly passive acceptance of uncertainty could lead to disabling consequences – perceived as self-defeating and intolerable fence-sitting – and even to suppressing the voices of experts who are courageous enough to make bold predictions that could prove in time to be correct. This is not the intention. Such bold opinions can stimulate useful debate. Certainly many individual experts might be well advised to give greater consideration to other possibilities before championing extravagant conclusions, and be prepared to adjust their views if circumstances should warrant; but such caution should not necessarily prevent them making bold predictions if they feel a sufficient degree of confidence that these are justified. However, decision-makers and other interested parties would also be wise to ensure that such individual views are considered along with alternative views. There remains a compelling case for an inclusive and durable research framework to test the persuasiveness of competing judgements over time.

Beyond dominance and collapse

Accepting and respecting the reality of major and continuing uncertainty about China's future prospects has several advantages. First, it offers a way to transcend a simplistic, sterile, divisive, and unproductive optimist–pessimist debate. Both optimistic and pessimistic scenarios are, in their extreme forms, equally questionable. The possibility that China might become dominant in its own region and even in the world, or that its multiple challenges might cause it to collapse, are both difficult to dismiss completely. Both are essentially plausible if based on generous assumptions. But mixed outcomes seem more likely. Many of the more extreme pessimistic or optimistic scenarios would require a rare combination of uniformly negative or uniformly positive influences. They appear equally unlikely to eventuate. It is perhaps more credible, or even likely, that China in the future will continue to be both strong and weak together – with still-growing influence at the same time as it struggles to overcome ongoing major challenges.

With regard to pessimistic scenarios, most would need to assume serious mismanagement and that expected counter-measures would prove ineffective. But incentives to adopt effective counter-measures would be enhanced, or accelerated, by the fact that both dominance and collapse would have very negative consequence for both China and the world. The idea that China will have the inclination and ability to achieve complete dominance, even in its own region let alone globally, seems far-fetched. It seems more likely that such dominance would be widely and probably successfully resisted if it went too far – in much the same way that the dominance of the United States, even as a unipolar superpower, has been widely resisted. China's dominance – whether politically, economically, culturally, or militarily – would arouse vigorous opposing forces and coalitions. It would also be completely contrary to its repeatedly declared intention to eschew hegemony, foster peaceful international relationships, and avoid interfering in the internal affairs of other countries.

What then of the prospects for China's collapse, another equally unappealing and destabilising outcome? Just as complete dominance seems inherently improbable, complete collapse is also a very questionable concept. The views of some pessimists who keep alive such prospects are admittedly not without some basis. The possibility of political collapse certainly cannot be ruled out. After all it has happened elsewhere, including in the case of the Soviet Union. But China's leadership appears very conscious of that scenario and determined to avoid its repetition. Given the apparent strength of their position and achievements and the vigilance of the internal security apparatus, we should perhaps not be too quick to anticipate the regime's early demise. Experience elsewhere has shown that change if it comes may come precipitately, but total collapse still seems unlikely. More probable could be a temporary economic slowdown, followed possibly by longer-term stagnation or even decline. But wholesale economic collapse seems likely to be averted if even only partly effective corrective measures were taken.

Excessively optimistic scenarios that see China's rapid rise continuing uninterrupted would also require that a questionable array of assumptions be realised. Some experts, including increasingly in China, appear to assume or be hopeful that its power will continue to grow until it becomes the number one country not only economically but also in a combination of other measures. But such optimism is offset by the growing number of others who argue persuasively that such ascendency would need to assume that China will be able to successfully address a lengthy and daunting list of vulnerabilities. Many of these experts see these vulnerabilities as so formidable that surely some combination of them must eventually drag down China's performance.[5]

No matter how eager experts are to produce forecasts about China that polarise in one direction or the other, uncertainty will be likely to continue to reign. But uncertainty, where it is real, is not without its advantages. In a situation where outcomes are not predetermined, there may be vitally important opportunities for appropriate intervention to shape events in new and positive directions. Rather than concentrating on the competing claims of equally problematic optimist/dominance and pessimist/collapse scenarios, a more credible and useful approach might be to construct two new scenarios by amalgamating, first, the potentially positive and, second, the potentially negative developments that might conceivably arise from the unfolding of the critical uncertainties examined above in each of the four domains. This would make it possible, for illustrative purposes, to paint either a very positive or a very negative portrait of China's future.

In a positive future China might, for example, permit the steady growth of political liberalisation at a pace consistent with internal stability, overcome vested interests and transition to a more equitable and durable economic model, accept fully the imperative of sustainable environmental development and act with courage and determination to accelerate the emergence of a low carbon economy, and successfully contain excessively nationalist influences to rededicate itself to a strategy centred on peaceful development and harmonious international relationships. In a negative future, at the other extreme, we might see a combination of a stubbornly authoritarian power that resists internal political reform, a declining economy unable to transition to a more viable development model, uncontrolled growth that overwhelms efforts to achieve environmental sustainability, and an imperious and xenophobic regime determined to pursue an expanding international agenda of inflexible core interests.

These polarised alternative outcomes are caricatures of two of the most extreme of China's possible futures, but neither is completely implausible. Each has its roots in present conditions. They serve to illustrate not just the frequently cited negative possibilities but also – and much more importantly – the scope for a combination of positive outcomes that would be enormously beneficial not just for China but for the world. These positive outcomes – many of which are already reflected in China's official policy formulations – are likely moreover to be eminently achievable if China and other countries are able to envision their feasibility and vigorously encourage their realisation. In this equation, a critical influence will be the beliefs and behaviour of the various national and sub-national actors within both China and other countries. Negative perceptions breed negative outcomes whereas constructive and positive perceptions, and a willingness to foster their actualisation, increase the chances of positive outcomes. In short, with skilful management and accommodation on all sides there seems no good reason to be unduly pessimistic about China's national development and international engagement, especially if appropriate attention is paid to embracing and shaping the essential uncertainty of future circumstances.

In such an environment, China experts will have a major and continuing role to play in alerting decision-makers and the public to potentially positive, as well as negative, outcomes. Experts can assist to identify not only likely futures but also normative futures that are not so much likely as possible and desirable. It may be useful to think in terms of *default* futures that arise from perceptions that the future is determined by fate or destiny, and *created* futures that can be promoted by proactive strategies that see destiny as a product of choice and human intervention. Default futures are likely to be realised when only one potential future is considered, while created futures may be realised through being mindful that alternative and more beneficial futures are achievable.

It is difficult to foresee how China's – and the world's – pressing economic, environmental and security problems can be addressed without such a reorientation of thinking. This is especially so, given that the problems of the twenty-first century seem very likely to be considerably more challenging than those of the twentieth. In this current century, it is increasingly persuasive that interdependence between countries is an inescapable fact, and that the ability of both experts and policy-makers to assess, manage, and exploit the continuing uncertainties of this interdependent environment will be decisive.

Notes

1 Kerbel, "Thinking Straight: Cognitive Bias in the US Debate About China".
2 Li, "The China Paradox and American Misperceptions", 27.

3 Oksenberg, Swaine, and Lynch, *The Chinese Future*, 26.
4 Cited by Andrew Moody in "Making head and tail about China", *China Daily European Weekly* (March 30–April 5, 2012). Perhaps the last word on this might be left to Mao Zedong, who in his essay "On Contradiction" observed that contradiction was "present in all processes" and that "the struggle of opposites is ceaseless". See Mao Zedong, "On Contradiction", in *Four Essays on Philosophy*, (Peking: Foreign Languages Press, 1968), 72.
5 In addition to sources cited in earlier chapters, see Jean-Luc Domenach, *China's Uncertain Future*, (New York: Columbia University Press, 2012); Timothy Beardsley, *Stumbling Giant: The Threats to China's Future*, (New Haven, CT: Yale University Press, 2013), 434–435; Mel Gurtov, *Will This Be China's Century? A Skeptic's View*, (London: Lynne Rienner, 2013); and Jonathan Fenby, *Will China Dominate the 21st Century?* (Oxford, UK: Polity, 2014).

Bibliography

Beardsley, Timothy. *Stumbling Giant: The Threats to China's Future*. New Haven, CT: Yale University Press, 2013.

Domenach, Jean-Luc. *China's Uncertain Future*. New York: Columbia University Press, 2012.

Fenby, Jonathan. *Will China Dominate the 21st Century?* Oxford, UK: Polity, 2014.

Gurtov, Mel. *With This Be China's Century? A Skeptic's View*. London: Lynne Rienner, 2013.

Index

For Product Safety Concerns and Information please contact our EU representative GPSR@taylorandfrancis.com
Taylor & Francis Verlag GmbH, Kaufingerstraße 24, 80331 München, Germany

www.ingramcontent.com/pod-product-compliance
Ingram Content Group UK Ltd.
Pitfield, Milton Keynes, MK11 3LW, UK
UKHW022130190926
475702UK00013B/1722
9780815371151